Saved by the Siesta

fight tiredness and boost
your health by unlocking
the science of napping

Brice Faraut

with Cédric Weis

Translated by Eric Rosencrantz

SCRIBE

Melbourne • London

Scribe Publications
2 John Street, Clerkenwell, London, WC1N 2ES, United Kingdom
18-20 Edward St, Brunswick, Victoria 3056, Australia
3754 Pleasant Ave, Suite 100, Minneapolis, Minnesota 55409, USA

First published in French as *Sauvés par la sieste* by Actes Sud 2019
Published in English by Scribe 2021

Text copyright © Actes Sud 2019
English language translation copyright © Eric Rosencrantz 2021

Typeset in Portrait by the publishers

Printed and bound in the UK by CPI Group (UK) Ltd, Croydon CR0 4YY

Scribe is committed to the sustainable use of natural resources and the use of
paper products made responsibly from those resources.

978 1 912854 72 1 (UK edition)
978 1 950354 57 3 (US edition)
978 1 925849 71 4 (Australian edition)
978 1 925938 41 8 (ebook)

Catalogue records for this book are available from the National Library of
Australia and the British Library.

scribepublications.co.uk
scribepublications.com
scribepublications.com.au

For my children, and for Fanny

Sachons dormir, nous saurons veiller.

By learning how to sleep,
we shall learn how to wake.

Alain (Émile-Auguste Chartier, 1868–1951)

Contents

Introduction   1

Part I   Portrait of the Sleeper
1. The Ins and Outs of Sleep   9
2. Healing Sleep   32

Part II   A World of Sleep Debt
3. The Sandman's Debtors   59
4. How Deep in Debt Are We?   76

Part III   The Siesta, or: The Force Awakens
5. Pocket Medicine   97
6. The Art of the Siesta   107
7. A Therapeutic Stroll   129

Conclusion   141
The Call of the Siesta: The Napper's Cheat Sheet   145
Acknowledgements   151
Bibliography   153
Notes   175

Introduction

Nowadays, instead of giving sleep its rightful place in our lives, we sacrifice it on the altar of work or subdue it with drugs, or at least food supplements. In my own country of France, sleeping pills are regularly taken by 6–7 per cent of the adult population and psychotropic drugs by 18 per cent, while we consume 1.4 million packs of supplements containing melatonin every year.[1] Millions of people grapple daily with a lack of sleep that saps their energy, undermines their health, and can even prove life-threatening. And yet there's a simple solution to the problem: get some sleep. Which, in our hyperactive society, often means *take a nap*. So how come doctors never prescribe this particular 'drug'? Why is this ancestral remedy so sorely underrated in our society?

It is, to be sure, a trending topic: the media increasingly relay the latest scientific findings on the benefits of the power nap, entrepreneurs tout its positive effects on productivity, and people look wistful at the very thought of it. Yet napping has hardly caught on. We're still liable to harbour doubts about the seriousness of daily practitioners of the afternoon snooze. We can't help suspecting it's a drag on productivity — even though 50–70 per cent of the nation's workforce complain about not getting enough sleep. Should we feel guilty about seeking to remedy an evil that's wearing us down? Who says we're allowed to sleep seven or eight hours a night — but not a few extra minutes every day? The answer is at least as old as the industrial world: despite the curative virtues of taking a snooze, many people still think it's a bit too easy, and a bit lazy, to give in to the temptation. On the contrary, however, there is wisdom in napping. In fact, it is of pressing importance to give the nap its due, to acknowledge its therapeutic benefits, and to reinstate it as a daily routine on a par with our regular meals, our daily grind, and our nightly repose. May this book serve to convince you of that.

And after all, the habit of napping isn't entirely alien to you: you did it every day, albeit sometimes reluctantly, for the first few years of your life. Yet while you may have renewed a nodding acquaintance with the occasional catnap on weekends or holidays, you've likely never dared to 'take it to next level'. But it's an ally for your daily life, and its aid is most effective when routinely enlisted. It's not just there to fill idle hours on lazy days of leisure, but to help your body and mind feel and function a little better every day, and to protect them from the chronic sleep debt from which all, or nearly all, of us suffer. And some more than others. For though sleep belongs to everyone, not everyone gets their fair share. Many in the world's labour force have to work nights, shifts, weekends, overtime. The bite that takes out of their nightly rest, estimated at an average of an hour and half per night for 20 per cent of the population, exacerbates the daily travails of the workaday world. So it's high time our leaders, in business and government, acknowledge and address the problem. Sleep is not lost money or time. Just as you can't have light without dark, life wouldn't be possible without sleep. (Isn't it said that we spend a third of our lives sleeping? If only that were true ...)

Everyone knows that sleep is 'refreshing' and 'restorative', but that's an understatement. Do you have any idea what's going on in your slumbering body? The function of sleep is not only to rest your muscles and reinvigorate you. It also enables growth, slows down the ageing process, consolidates useful information and disposes of useless clutter in your memory, synthesises the proteins necessary for the upkeep and regeneration of your cells, eliminates toxins, reorganises neural connections, boosts your immune system, fuels creativity, maintains wellbeing, and preserves your equilibrium. This is why it's so important to protect your nightly rest from everything in our ultra-connected day and age that might adversely affect it — especially the blue light of electronic devices — combined, whenever possible, with its natural daytime counterpart, its de facto scale model: a restorative nap. Chronic sleep deprivation takes a toll on our mental health, which is no surprise: losing our sleep is losing part of our selves. To remedy the situation, some take sleeping pills or antidepressants, but at the risk of meddling with the architecture of sleep, altering its primary functions and causing irreversible long-term damage.

Studies conducted all over the world — including the research my colleagues and I coordinate at the Hôtel-Dieu Hospital in Paris on the functions of sleep, and especially on the nap — show that napping is a highly adjustable all-purpose remedy for sleep deficit, a 'medicine' for the future, many of whose virtues are now known to us. We've come a long way since the firmly held Age of Enlightenment notions that sleep involves a 'compression of the brain', that snoring is 'the sign of deep sleep', and that 'only illness and extreme heat can induce men to take a nap'.[2] Science has progressed, and in recent years a considerable number of studies have shown how napping can not only cure extreme fatigue, but also combat drowsiness, pain, gloominess, immunological fragility, stress, hypertension, obesity, and cardiovascular disease. How many of us would simply live better, and longer, if we paid more attention to the benefits of sleeping better, or more, and above all more frequently?

At least one great man is said to have set an example for us: a certain Leonardo da Vinci, who treated himself to a good 15-to-20-minute snooze every four hours. He apparently grasped, long before his contemporaries (as in so many other domains), that sleeping and snoozing are conducive to genius. Even if this is just another legend, it is not without solid foundation: napping does indeed have the effect of a marvellous magic potion on human physiology. To reap the rewards of this practice, however, we need to know the finer points: the right sleeping positions, the times of day conducive to falling asleep, the most effective durations, the stages of sleep that boost alertness, cognitive performance, memory, and creativity, and tips on how to fall asleep quickly and wake up without feeling drowsy.

No one has ever survived prolonged total sleep deprivation, and the most tenacious (or craziest) of them all, the legendary Randy Gardner, only held out for 264 hours, or 11 days. The experiment took place in the United States back in 1964 and was never repeated. The 17-year-old volunteer guinea pig reportedly suffered moodiness and coordination problems after three days without sleep, and hallucinations by the end of the fifth day. His memory was malfunctioning by the seventh day, and he was seized with paranoia on the tenth day. On the last day, he was dazed, his face drained of all expression, and he asked to stop the experiment: he no longer understood what it was about.[3]

Five years earlier, likewise in the US, a disc jockey by the name of Peter Tripp survived 201 hours without sleep in a 'wakeathon' to raise funds for the March of Dimes charity. He, too, started hallucinating after four days: the controls on his mixing desk morphed into columns of swarming insects, and the physician tasked with helping him overcome his anxieties became an executioner hell-bent on burning him at the stake.[4] Since then, other experiments, better prepared and less stressful, have borne out the biological utility of sleep. Most of them are carried out in the laboratory and controlled by sophisticated equipment that uses electrodes to measure the activity of the subject's brain, muscles, and eyes, as well as blood and saliva samples to monitor the ever-fluctuating composition of body fluids. The lay reader is likely to be as surprised as many a scientist at how wonderfully active our slumbers are despite their apparent serenity.

But let's begin at the beginning. Before delving into the sleeper's inner world, let's attempt a rough portrait of the sleeper to see how we sleep, how much sleep we need, and what our nocturnal habits and rhythms are. This will give us a better understanding of the tremendous machinery that sleep sets in motion inside the body and brain. These tools will give us a better understanding of the benefits of napping, its techniques and therapeutic potential, and why it's a real public-health issue in our modern-day, sleep-deprived world.

Part I

Portrait of the Sleeper

1

The Ins and Outs of Sleep

All sleepers look a bit alike, and yet we'd be hard put to find any two people who sleep the same way. Naturally, we often see them lying on their backs, bellies, or sides, lips parted, eyes gently shut, faces softened or sullen. Sometimes their arms move about and they seem surprised by their body's caprices. They smile in their dreams, frown in their nightmares, and eventually stretch their limbs with obvious relish. But however similar they may be, no two sleepers have the same way of beginning their nightly repose and dividing it up, making it last, interrupting it, and emerging from their slumbers sated and reinvigorated the next morning. This is because, for one thing, sleep needs differ according to numerous criteria, not the least of which is age. Our sleeping habits are also partly determined by our genetic heritage. Nevertheless, in the east and west, north and south, whether the days are longer or shorter, whether the sun sets early or late or shines nonstop, a day lasts only 24 hours.

The human clock, our inner sun

AN OBSESSION WITH RHYTHM

Since the dawn of time, the fall of night has inexorably summoned the animate world to sleep. But what law decrees that nature, animal and vegetable alike, should suddenly nod off when the sun goes down? It is because of a marvel of miniaturisation, at least in the mammalian brain, that enables the body to measure time. The body's 'master clock' — or 'central pacemaker', as it's also called — comprises two sets of 10,000 neurons, each the size of a pinhead, which scientists have unflinchingly dubbed the 'suprachiasmatic nuclei' of the brain's hypothalamus.[1] Its job is to control a great many biological rhythms that synchronise body functions over each 24-hour period — hence the term 'circadian' (from the Latin *circa*, 'approximately'; and *diem*, 'day'). Those rhythms include not only alternation between sleep–wake cycles, but also our body temperature, the release of hormones, our respiratory and heart rates, blood pressure, digestive functions, and all our nutrient and energy metabolism, without which we wouldn't be alive.[2] Even our physical and intellectual performance depends on this inner ticking, which has recently been shown to play a crucial role in cell division and the repair of our DNA as well.[3] Even in our earliest infancy, this biological clock puts a stop to our wriggling and cooing in the cradle every four hours and makes us drop off again. And when we grow up, it makes our eyelids grow heavy at just about the same time just about every day, and makes us wish — without admitting it to a soul — we could simply lay our head down on the desk in front of us for a little snooze.

The circadian rhythms of several body functions controlled by our biological clock.

TO EACH OUR OWN TEMPO

Even if it seems to be directly synchronised with daylight, our biological clock is autonomous. Put a *Mimosa pudica* (aka 'sensitive' or 'sleepy' plant) in an opaque box and its leaves will reopen at dawn all the same, and every bit as proudly as outside in a garden.[4] Hole up for several months in a flat without any light and your biological clock will set your daily pace all the same — according to your particular cycle, for everyone has their own particular cycle. The average cycle takes 24 hours and 12 minutes, but may be shorter (ranging down to about 23 hours and 30 minutes) or longer (up to about 24 hours and 30 minutes). It all depends on how fast your clock runs. For example, according to a 2011 American study of 157 subjects by a team from the Division of Sleep Medicine at Harvard Medical School, women's clocks run faster, which would explain why their first 'sleep window' opens 30–45 minutes before men's,[5] and why they're more likely to be early risers, as suggested by a previous study of 55,000 subjects by the University of Munich's Centre for Chronobiology.[6] How much tension could be defused, especially in couples, if only we were more aware of the subtleties of our biological clocks!

So you're more likely to be an early bird if you have a speedy clock and a night owl if your clock tends to take its time. This is called your 'chronotype', and it has an impact, as we shall see, on the ideal timing for a nap. Our individual chronotype also changes as we age. So we needn't necessarily berate our teenagers for going to bed later and later: it may be due, at least in part, to the natural development of their chronotype.[7] This shift is, by the way, far more pronounced among males than females: women go to bed latest at the age of 19.5, on average, and men at 21; men also tend to turn in later than women throughout adulthood until their 50s.

FOLLOW THE BATON

In any case, in order for our master clock to remain in sync with the solar day, it has to be reset every day. And for that purpose, the suprachiasmatic nuclei can rely on their lookout posts — the retinas of our eyes — to keep them constantly updated on the ambient luminosity. So bear that in mind as night falls, because interior lighting and LED screens significantly affect our biological clocks. First of all, while our clocks may not be very sensitive to light in the late afternoon, they're a lot more sensitive around the usual bedtime hours, namely 10.00–11.00 pm.[8] Secondly, light at night markedly inhibits the secretion of melatonin, the hormone responsible for switching the body's activity to night mode and leading us off to the Land of Nod. Thirdly, blue light, which electronics manufacturers use in most display screens for both aesthetic and economic reasons, desynchronises our biological clock as much as a white fluorescent light a hundred times more intense.[9] Consequently, blue light at night, however dim, tends to throw our circadian rhythms out of whack,[10] so taking one last peek at the internet on your computer or phone right before going to bed is the very worst of late-night habits. In a word, there's nothing like deep soothing darkness (at night) and bright sunshine (in the morning) to set our clocks right and help us get out of bed on the right side in the morning.

Our master clock is like a conductor keeping time to ensure that all our physiological functions come in on cue and play their parts in harmony, 24 hours a day. Without resetting your clock daily, these functions would naturally occur about 12 minutes later with each passing day, with various consequences in the long run: a loss of appetite at lunchtime, a sudden late-afternoon chill, a desire to dance around the living room in the middle of the night, and a bad case of the munchies upon waking from a dream. As French neurobiologist Claude Gronfier puts it: 'Without the appropriate light hygiene, our clock loses the beat, and the result is cacophony.'[11] Jet lag gives us some idea of the effects: sleep, alertness, mental acuity, and memory are immediately altered after, say, flying across several time zones and landing in China. But we're lucky if that only happens to us once in a while. Many people suffer from 'chronobiological' disorders all the time, and without the saving grace of a change of scenery: those who are totally blind[12] or stricken with Parkinson's or Alzheimer's disease[13] or certain mental illnesses, but also more and more night and shift workers. They all inevitably suffer from sleep disorders, decreased alertness, and memory impairment, as well as alterations of their cardiovascular and immune systems. Worse still, according to a 2007 study by the International Agency for Research on Cancer (IARC), the impact of clock desynchronisation on the body's cellular and molecular functions may be responsible for the increased prevalence of breast cancer among women who work at night.[14]

KEEP ONE STEP AHEAD

Meanwhile, the sun has gone down and, as the hour for us to follow suit approaches, we should enact a bedtime ritual. As you may have noticed, conductors have mastered the art of anticipating the beat. And just as the strings in the orchestra ready their bows to attack the next note right on time, we'd do well to ready ours — especially if we tend to have a hard time falling asleep. We may take our cue from melatonin: every night, this 'time-telling' hormone makes its way through our bodies two hours before we drop off, as if to warn us that daylight is fading and the night is nigh. So why not do our part by gradually reducing the ambient light? And since the body needs to lower its internal temperature slightly in order to fall asleep — after peaking at 37.5 °C around 5.00 or 6.00 pm, it bottoms out at 36.5 °C around 4.00 am — why not help it cool down by taking a lukewarm shower before slipping between the sheets (in the morning, conversely, a hot shower will help wake your body up). Mental preparation helps, too, so think 'sleep' two hours before turning in. This is precisely what your master clock and all the physiological functions it coordinates do every night. Cast off any nagging thoughts that keep you from falling asleep and sleeping at all by scribbling them on a piece of paper: that way, you'll know they're there, waiting for you, and you won't be afraid of forgetting them. And if need be, make arrangements to ease your awakening, too: get hold of a special lamp that will gradually steep your bedroom in early-morning light. Here again, your biological clock does its part by producing a stress hormone two hours before your eyes open: the famed cortisol, whose small vices and great virtues will be considered later. All these physiological preparations are thrown into disarray by jet lag, for example: it takes eight days for our body temperature, and two to three weeks for our secretions of melatonin and cortisol, to get back on track.

CELLULAR TICK-TOCK

Up to this point, we've been discussing the master clock as if it alone could coordinate the many physiological processes involved in the sleep–wake cycle. But this fails to take into account the ingenious complexity of our bodies, which have resourcefully slipped into each of our cells tiny so-called 'peripheral' clocks, which act in concert with the master biological clock according to the role assigned to their host cell. As a result, the body may be more sensitive to certain drugs depending on the time of day. Lidocaine, for instance, is an anaesthetic used by dentists whose effectiveness varies according to the time of day. The body is most receptive to its anaesthetic effects around 3.00 pm and least receptive around 10.00–11.00 am or 7.00 pm, so it's best to administer the drug accordingly.[15] Our bronchial tubes, on the other hand, have a circadian rhythm that narrows their diameter from 9.00 pm to 5.00 am, which is why asthma attacks usually occur at night.[16] Another example: the peripheral clocks of the liver and pancreas are generally in sync with the body's biological clock — the resting period (at night, for humans) corresponds to a fasting phase. But if you feed rodents during their usual resting phase (daytime) for a week, you'll find that the peripheral clocks of their liver and pancreas will decouple from the master clock to recognise a sole authority: meal time.[17]

The primacy of the master clock is due to the fact that it's the only one connected to our retina and therefore directly attuned to daylight. But sleep also heeds the ticking of lesser-known clocks. One such clock controls the alternation of our nasal cycle and, much to the detriment of the quality of our sleep and for reasons long unknown to us, makes us roll over every 60–90 minutes from one side of our bodies to the other, namely the side with the currently congested nostril. A nap, however, is usually brief enough to be spared such tossing and turning, and, as we shall see, its therapeutic effects serve to repair some of the damage caused by the frequent desynchronisation of our master clock.

Night cycles: the stages of sleep

A THREE-STEP WALTZ

Unless disturbed by insomnia, our nights usually leave us with a sense of unity. And the best nights are those deemed complete and whole, i.e. uninterrupted. But even a good sleep is not a uniform whole: on the contrary, it is broken up into a series of four to six cycles interspersed with awakenings — imperceptible awakenings that are almost never remembered, provided they don't exceed a few minutes' duration. And in each of these cycles, three types of sleep follow one after another continuously, as in a relay race: light slow-wave sleep, deep slow-wave sleep, and REM sleep. Taken as a whole, sleep thus resembles a ride on a local train that stops at four to six stations (cycles) every night, the exact nature of this journey determined by the sleeper, their nights, and their naps. The image of a little train comes naturally to my mind, by the way, when specialists set about describing sleep. Picture it: an old-fashioned train with carriages connected by a gangway.

THE NIGHT TRAIN

Before going into detail, why not take the analogy a step further by watching yourself sleep on the little train? As a matter of principle, when you catch the sleep train, you sit in the first carriage, the light slow-wave sleep carriage. Soon enough, though, you can't resist the urge to move to the second carriage, the one for deep slow-wave sleep: there's no denying that it's restful and, as your slumbers begin, you always pay special tribute to it. However, since all good things must come to an end, you move on. The third carriage, which is the last one in the first section of the train, is for REM sleep. Oddly enough, it lends itself better than the others to dreaming. Bear in mind that, at the beginning of the night, your stay here never lasts long.

So far, everything has gone without a hitch. But suddenly the train stops. Which wakes you up, naturally, as if you were saying to yourself, 'It's now or never. Cross the gangway connection and try the second section of the train, even if it means turning my back on my first dreams.' So off you go again: seated in the second section, you fall back asleep before you know it in the light-sleep carriage. Making the most of each period differently, you go through each of the three carriages in the new section: light slow-wave sleep, deep slow-wave sleep, REM sleep. At the end of the second section, same thing again, another halt — you wake with a start again and shuffle along to the third section of the train, which you move through at a different tempo: light slow-wave sleep (getting longer and longer), deep slow-wave sleep (getting shorter and shorter), REM sleep (getting longer and longer). This process is repeated continually till you reach the last stop, i.e. at the very end of the train, meaning the end of your night's slumbers. How many sections you move through per night depends on your sleeping habits.

THE MECHANICS OF BRAINWAVES

Overall, you spend on average 40–50 per cent of your time in light sleep, 20–25 per cent in deep sleep, 20–25 per cent in REM sleep, and 5–10 per cent awake. But how did you manage to nod off so smoothly in the first carriage? What happened in your brain the instant you shut your eyes? Well, what happens in a theatre when the curtain falls: the tension subsides, the world of live performance stops, the audience slowly returns to their lives, and the bare stage, nestled in darkness, falls asleep. When your eyelids fall shut, the wonderful world of brainwaves is slowed down by a complex system that inhibits waking systems and by the accumulation of adenosine, a molecule produced by the breakdown of ATP, the energy-carrying molecule that our bodies consume all day long.[18] When you're awake with your eyes open, caught up in the spectacle of your life, you're functioning at full capacity and your brainwaves have a very high frequency of over 13 Hz (13 cycles per second), known as beta rhythm. But you need only close your eyes to relax and the frequency will drop immediately down to between 8 and 13 Hz (alpha rhythm). Open your eyes again and you're back in beta rhythm. But if you lie with your eyes shut in the dark, your brain activity will shift down from alpha to theta rhythm: 3–8 Hz. That's when sleep sets in and you find yourself in the first carriage of the train ...

At this point, your body will gradually adapt to the new rhythm. This is stage one of sleep. Under your closed eyelids, slow eye movement can be observed. Your muscles relax, but not yet completely, and your breathing and heart rate slow down.

You're still sensitive to noise, and any disturbance is liable to wake you up. Some dreamlike images immediately appear in your mind's eye. If nothing wakes you, you continue on to stage two: you haven't left the first carriage yet, but your sleep is getting heavier, with less eye movement. Your muscles, while increasingly relaxed, still have some tone; your brain activity, though markedly slower, still exhibits 'sleep spindles', bursts of high-frequency (12–14 Hz) brainwaves usually lasting 0.5–1.5 seconds; and your breathing is still deep and irregular. When the theta rhythm finally reaches its lowest level, you've reached the end of the first carriage, the end of light slow-wave sleep.

Now you can make your way into the sumptuous second carriage. This is stage three, that of deep slow-wave sleep. Your brain is now submerged in slow large-amplitude waves of 1–3 Hz (delta rhythm), and your breathing finally becomes regular and of small amplitude. You can no longer be woken up unless someone shakes you or shouts in your ear.

SLUGGISHNESS AND NEWFOUND VIGOUR

Once we know that light sleep is shortest and deep sleep longest during the first part of the night, we can understand why midnight has come to be known as the 'witching hour'.[19] Yet it is during this 'perilous' abandonment that your body is able to fully recover (we'll get back to this point later on). Then, all of a sudden — usually after 70–90 minutes on the train — the long slow waves give way to a sudden increase in brain activity: you've just entered the third carriage. The brain's electrical activity here is no longer very different from being awake — as are your respiratory and heart rates, which increase and become irregular again, and your eyes, which move fitfully under your eyelids. Your muscles, on the other hand, no longer have any tone, leaving you all but paralysed. In short, you show signs of both wakefulness and deep sleep. Paradoxical, isn't it? Which is why this final stage, called REM (rapid eye movement) sleep in English, is called 'paradoxical' (or 'stage P') by the French, who love contradictions.[20] This is the only stage readily recognisable without any sophisticated equipment because — and this goes to show what strange energies are unleashed during REM sleep — it's always accompanied by penile erections in healthy men and unusual uterine contractions of greater amplitude in women.[21] An 'oddity' observable in other mammals, too, and about whose origin you must have often wondered when, woken up by the stopping of the train, you were surprised to discover this erotic tension. Whatever the dreams that so mysteriously fill your REM sleep, this physiological phenomenon is merely a consequence of the somewhat anarchic workings of your nervous system during this particular stage, and not an expression of your libido.[22]

Sleep at all ages, or: the gradual diminution of repose

At first glance, the stages of sleep seem to be shared by just about all the other warm-blooded creatures on the planet. When they close their eyes, an African monkey, an Australian kangaroo, and an American mouse don't seem all that different from us: they cut themselves off from the world and drift off on alpha, theta, and delta waves, their temperature drops a little, and they roll their eyes, dreaming of what may come. But we'd be wrong to see this as a form of equality, since what matters most is the overall duration of sleep in relation to the need for it, and its healing power, which depends on how much time is spent in each stage of sleep. In these respects, sleepers vary considerably. Before looking into what sets them apart, let's see which factor tends to bring them together: their age. Because sleep requirements are not the same for all generations: they preponderate in the pram and pervade childhood, let up a bit during adolescence, stabilise in adulthood, and eventually peter out, at which point we bid goodbye to the long restorative and revitalising nights that once gave us the impression of being fully rested.

CHILD'S PLAY

Let's begin at the beginning. In your infancy, you slept like a house cat, 16–17 hours a day, and resurfaced, just like a cat, whenever your appetite resurfaced, every three or four hours.[23] Up to three months of age, your sleep was still in its infancy, too, alternating between quiet sleep (soon to become your slow-wave sleep) and restless sleep (your future REM sleep). From 6–12 months of age, you got accustomed to sleeping 10–11 hours at night with 3–4 hours of scattered naps during the day. Your nights were marvellously rich in deep slow-wave sleep (30–40 per cent) and REM sleep (30 per cent).[24] And they'd have stayed that way for a few more years had it not been for primary school, when classes too early in the morning began encroaching on the 10 or 11 hours of night sleep you still needed, depriving you of part of your REM sleep — which happens to be essential to your ability to learn and memorise and to your psychological wellbeing.[25] This loss was all the more detrimental since your REM sleep was gradually decreasing during childhood anyway,[26] as was your overall sleep, from 12 hours (between three and five years of age) to 10 hours (between 10 and 12 years of age).[27] Fortunately, your deep slow-wave sleep, which is crucial for growth and, as noted above, most important at the beginning of the night, remained stable (at roughly 30 per cent).[28] So you didn't do too badly after all.

THE FIRST CRISES

As your need for sleep decreased in adolescence, you were able to settle for nine to ten hours a night until adulthood. Though we still had to keep an eye on you, because you had only one desire at the time: to stay up later and later, to isolate yourself, to question everything, to communicate, to resolve a thousand personal problems in secret. Your evening habits sometimes made it hard to concentrate in class and might even have made you drowsy during the day. To be sure, your sleep was still adaptable enough for you to recoup on weekends the hours of sleep sacrificed during the week. This trick was but a poor remedy, however: for one thing, it didn't guarantee complete physical recovery; for another, it imposed an additional rhythm on your body, a sort of 'artificial jet lag' that insidiously uncoupled it from its circadian rhythm. In any case, that luxury eventually disappeared.

And now you're grown up. Your sleep has taken on a more personal, but also more rigid, configuration, and its stability has come to depend on the patterns of your social and professional life.

THE GRADUAL DECLINE OF SLEEP

Deep slow-wave sleep accounts for only 20–25 per cent of your adult nightly slumbers, as does REM sleep, and you wake up at night, however briefly, more frequently. That's the way life goes: over the years, sleep undergoes a gradual decline, becoming lighter and more unstable, interspersed with more and longer awakenings, especially during the second half of the night. As a result, we often fail to feel fully refreshed and restored in the morning. Deep slow-wave sleep gradually diminishes around the age of 40 and becomes increasingly rare after 70.

So it stands to reason that daytime tiredness should become more common over the years. Is there anything we can do about it? Can we replace deep slow-wave sleep with more REM sleep or more light slow-wave sleep? Given the functions of sleep, as we shall soon see, we shouldn't pin our hopes on that solution. Taking naps, on the other hand, can be quite effective in helping to offset the deleterious effects of gradually losing deep sleep as we grow older. First, however, let's look at how sleep requirements vary from one person to another, regardless of age.

Short and long sleepers:
a history of privilege

What could bespeak the importance of sleep more com-
pellingly than the fact that we spend about a third of our
lives asleep? Yet a rapid calculation based on a 24-hour day
immediately calls that contention into question: who among
us sleeps exactly eight hours a day? Not very many people,
by the looks of it. And not only, as one might suppose, for
professional or family reasons. These are clearly a factor and
we'll get to them. But what interests me at this point is our
physiological need for sleep irrespective of the constraints
imposed on us by our day-to-day lives.

THE MORPHOLOGY OF SLEEP

There are three types of sleepers: 'short', 'medium', and 'big'.
'Short sleepers' are people who need less than five or six
hours of sleep per 24-hour period to remain alert during the
day. 'Medium sleepers' need seven to eight hours a day, and
'big sleepers' at least nine hours. But it's hard to estimate
the percentage of each type among the overall population
using the only means at our disposal, the epidemiological
questionnaire. Because even if respondents can tell us how
much they sleep (with the margin of error and imprecision
inherent in this type of study), that may have less to do with
their actual needs than with their various constraints: work
schedule, family life, illness, sleep disorders, etc. All the same,
it's estimated that true 'short' and 'long' sleepers make up
roughly 7 and 10 per cent of the population, respectively.[29]

As we shall see presently, many employees, schoolchildren, and college students who call themselves 'short sleepers' are in fact sleep-deprived. For the sake of their own health, they shouldn't try to compete with the 'featherweights', who make do with four or five hours of sleep a night and — as if that weren't enough — are very active from the crack of dawn, if not earlier. There's no denying the obvious advantages of leading such an active life; however, an epidemiological study conducted by the department of psychiatry at the University of California, San Diego between 1982 and 1988 among a very large sample of the population (over 1.1 million Americans aged 30-102) found that people who slept an average of seven hours a night had the lowest mortality rate, and those who got by on only four hours or exceeded eight hours and 30 minutes a night had the highest death rate.[30] Then again, the ideal number of hours of sleep for the human body can't be deduced from this study, because it was based on the respondents' actual hours of sleep and not on their real needs.

GENDERED SLEEP

Who sleeps more, men or women? If it's difficult to study the real sleep requirements of a large population, it's no easier to break them down by gender. In any case, the total hours of sleep people report will always be skewed by a bias: namely, subjectivity. Some respondents will underreport the number of hours they sleep due to failure to keep a rigorous count or a sort of misplaced pride. Hence the value of laboratory studies to corroborate findings gleaned from questionnaires. But telephone surveys are so much easier to organise, and increasing the number of surveys is an efficient way of consolidating the data collected. Not only that, but they have the virtue of being mixed: as Belgian psychologist Myriam Kerkhofs points out, laboratory studies have 'essentially focused on men's sleep' on the grounds that women's hormonal fluctuations — accompanying their menstrual cycles and periods of pregnancy — significantly impact their wellbeing, health, and sleep.[31]

While mixed studies in the lab are more frequent today, they have not conclusively shown any significant difference between the sexes in terms of actual sleep needs. On the other hand, a vast epidemiological study conducted in 2014, involving 13,709 women and 10,962 men aged 15–85, by the Hôtel-Dieu Sleep Centre showed more men than women among the self-reported shortest sleepers (+33 per cent and +20 per cent in the categories reporting 4–6 and 6–7 hours of sleep, respectively), and more women than men among the self-reported longest sleepers (+25 per cent and +33 per cent in the categories reporting 8–9 and 9–10 hours of sleep, respectively).[32] That same year, the department of psychiatry at the University of Pennsylvania School of Medicine published a study of 34,320 women and 26,567 men aged 15 and up, which came to a similar conclusion: across all ages, men reported on average about ten minutes less sleep than women, but this gap, as the French study had found, narrowed as women passed the age of menopause and even more significantly as the respondents approached the age of 60.[33] Needless to say, these conclusions merely indicate average behaviours, and there are women in every category, including that of short sleepers.

THE PROGRAMMED ANIMAL

Whatever the type of sleeper you are, it is our tireless conductor — our biological clock — that runs the show. This is true to such an extent that, even if deprived of sleep for 24 hours, short and long sleepers alike will find that their levels of sleepiness closely correspond to their habitual sleeping time, as regulated by the release of melatonin and the lowering of body temperature before falling asleep and the production of cortisol before waking up. This experiment was performed in a 2003 study coordinated by the National Institute of Mental Health in Maryland. After being kept awake for 24 hours, short sleepers felt the first onset of drowsiness at 11.00 pm (two hours before their usual bedtime) and gradually experienced the effects of fatigue until 6.00 am (half an hour before their usual wake-up time); long sleepers also felt the first effects of drowsiness at 11.00 pm (the exact hour of their usual bedtime), and these effects intensified until 8.30 am (half an hour before their usual wake-up time). In both cases, the subjects — deprived of sleep for the benefit of science — found that their state of sleepiness decreased at their usual wake-up time, i.e. the precise time at which their cortisol habitually kicked in.[34] Two years earlier, the same team of researchers undertook an even more extreme experiment, which involved keeping subjects awake for 37 hours: less sleepiness was observed among short sleepers during the final ten hours of their sleep deprivation — proof of their marked capacity to resist the pull of sleep.[35]

LIKE MOTHER, LIKE DAUGHTER?

Why these differences? How come we can't choose our sleep needs any more than we can the colour of our eyes? Is one born a short, medium, or long sleeper? Does genetic determinism make us early birds or night owls? This is what genetics research is trying to find out, inspired by the fact that similar circadian rhythms are not infrequently observed in members of the same family. Obviously, this isn't sufficient proof, but it is sufficient grounds to ask questions. Identical twins — seeing as they have the same genetic heritage — were the first to provide some answers. Their chronotypes and sleep stages have been compared in a great many laboratory studies since the late 1970s: the takeaway is that genetic inheritance is about 50 per cent responsible for how long we sleep.[36] So studying twins can't tell us the whole story; we need access to a larger sample of the population. And we need to find out the real sleep needs of a very large number of people irrespective of environmental factors and subjective assessments.[37] Unfortunately, these difficulties have proved insurmountable. The most significant study to attempt to overcome all confounding factors mustered 4,251 subjects in 2013, but its results could only minimally (5 per cent) attribute how much we sleep to our genetic makeup.[38]

Researching genetic predisposition poses another problem: without manipulating our genes, which generally involves removing one particular gene or another to observe the consequences, there is, at present, no way of gaining an understanding of their genetic functions. So volunteers are hard to come by. This is why the fruit fly (*Drosophila melanogaster*) — also known as the 'vinegar fly' — became the guinea pig of choice for geneticists in the 20th century. In the early 1990s, scientists succeeded in isolating several genes in the fruit fly's DNA, mutations in which could 'eliminate or considerably shorten or lengthen circadian rhythmicity'.[39] This research, which was eventually rewarded with a Nobel Prize in Physiology or Medicine in 2017, did not explain the genetic nature of sleep durations either, but it did reveal the influence of certain genes on animal chronotypes. Deprived of any environmental cues, the insects fell asleep in an entirely erratic manner. This was an interesting discovery, seeing as we share 70 per cent of our genes with this insect — though that won't help us fly. And the discovery was soon put to good use in an animal even closer to us (90 per cent shared genome): the mouse, in which everything tallied, with the same circadian-rhythm genes precisely determining the animal's chronotype.

Then it was our turn. Although studies have been confined to collecting saliva from volunteers (in order to analyse DNA and then associate variants of our genes with particular sleep habits), the results were no less interesting: in 2010, a sleep specialist and a molecular biologist from the University of Surrey hypothesised that two variants of a particular biological-clock gene might be involved in establishing our chronotype.[40] As for the genetic determinism of how much sleep we need, the studies on fruit flies and mice seem convincing for the time being; those on humans are still too recent to allow us to set out their ramifications here.[41] In 2009, to be sure, a study of a family of short sleepers revealed that the mutation of one of the genes regulating the genetic expression of their biological clocks sufficed by itself to account for the brevity (six hours) of each family member's sleep.[42] And that was corroborated in 2014 by another family of short sleepers affected by a mutation of the same gene.[43] But the research is still in its infancy and, in this particularly complicated field, we must be patient. Eventually we'll know a whole lot more ...

In the meantime, let the flies fly, let the mice scamper freely, and let's get a closer look at what sleep is for in the first place.

2

Healing Sleep

The contrast at night between an inert body and the seething agitation inside it is so striking that it's hard to imagine anyone could possibly sleep in the midst of such a commotion. Everything outside seems deserted, but as we delve inside we discover a world readying for battle, with a whole throng of medical teams, as biddable as worker bees in a hive, preparing to invade every nook and cranny of the body to carry out repairs and remedy all the damage done during the day.

We all know how important it is to maintain complicated machinery and how regular upkeep guarantees its longevity. Well, our bodies are no exception. But the workings of mammalian bodies and brains are so energy-intensive, so frenetically active from morning till night, that they need to be cooled off, overhauled, and restored on a daily basis. This alone suffices to explain why we don't go to every party we're invited to and why, at nightfall, we generally heed the call of sleep instead.

So here we go. Let's start by lying down in bed. Fortunately, the medical teams are concealed inside your body — otherwise, your bed would look like an operating table. Now close your eyes. Your muscles relax, your tension subsides, your brainwaves slow down. That's it, now you're dozing off. Certain molecules in your blood, however, are already having a field day.

The functions of sleep

THE HORMONES THAT CONTROL OUR BODIES

Our hormonal system, composed mainly of a set of glands arranged vertically along the vertebral axis, secretes chemicals that circulate in the blood, delivering messages to cells in order to stimulate or inhibit their actions. This chemical control over the vital functions of our cells ensures our nutrition, breathing, digestion, growth — in a word, our survival. The endocrine system (from the Greek *endo*, meaning 'inside', and *krinein*, 'to secrete') irrigates the whole body 24 hours a day. It may be more active at certain times of the day, however: for one thing, because our biological clock makes it secrete many of its chemical messengers cyclically; for another, because it sometimes needs a good night's sleep to fully express itself.

We have about 50 different hormones, each of which has a well-defined role to play: some are specialised in cellular metabolism; some stimulate growth in children and adolescents and slow the ageing process in adults; some respond to countless episodes of everyday stress; some are involved in reproductive processes; some control our appetite and regulate our blood-sugar levels; and so on. Though they may prefer to wait till we're sleeping, studies show that many of them have (totally or partially) broken free from the dictates of our biological clock, taking action regardless of the hour at which we close our eyes to sleep. This helps to explain why sleep deprivation, from the moment it begins to directly affect the endocrine system, can cause illness. We don't sleep just to wake up alert and in good spirits: we sleep in order to grow and develop (as children) and to restore our bodies (as children and as adults).

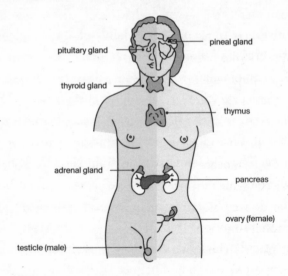

The major glands of the endocrine system.

SLEEP-DEPRIVATION EXPERIMENTS

Sometimes the best way for researchers to find out what goes on during sleep is to deprive us of it. All night, if possible. This is sufficient to wreak some havoc and find out whether certain metabolic processes that are essential to our bodies don't depend on sleep for their cues. And this is how, for half a century now, we have verified that certain mechanisms involved in the upkeep and regeneration of our bodies are triggered only in our sleep.

Let's take a group of 18-to-35-year-old men, isolate them in a laboratory and, for starters, continuously record their sleep by means of polysomnography (which records brain activity and that of the chin and eye muscles).[1] In order to do more than merely analyse the behaviour of their brains, let's attach catheters to their arms and take a little blood every half-hour. In order not to disrupt their sleep, let's set up our analytical equipment in an adjacent room and put at least one member of the research team there. The next night, let's keep the subjects from getting any shut-eye: they can read, watch TV, listen to music, observe the sky, chat with and stimulate one another to keep from nodding off — while we carry out the same measures as the previous night.

What do we observe from one night to the next? That all hormonal functions *dependent on the biological clock* take place at exactly the same time, regardless of whether the subjects sleep or not. This is perfectly normal since their clocks have not had time to adjust yet (remember that it takes several days to recover from jet lag and adjust our biological clock). But other hormonal functions seem to *depend entirely on sleep*. In our experiment, these hormones didn't completely disappear during sleep deprivation — since hormone production continues uninterrupted, though scaled down — but their secretion did not peak as usual during slow-wave sleep. This was the case with one of the most important hormones in the endocrine system:

growth hormone, which is essential for development in children and adolescents, cell regeneration in adults, cartilage maintenance, and protein synthesis. For growth hormone, sleep means action.

THE GOOD VIBES OF GROWTH

For half a century now, scientists have suspected that a massive release of growth hormone takes place as soon as we enter the first carriage of deep slow-wave sleep, no matter when we first fall asleep.[2] In 1992 and again in the year 2000, studies by Eve Van Cauter, a Belgian professor in the department of medicine at the University of Chicago, established more formally the causal link between leaving deep slow-wave sleep and slowing down growth-hormone production.[3] A 2004 study by her team showed that lengthening this stage of sleep — which was done deliberately, using a psychotropic drug — proportionally increased growth-hormone secretion.[4] And six years later, German neurobiologist Jan Born's team at the University of Lübeck hypothesised that a high concentration of growth hormone during deep slow-wave sleep could strengthen immunological memory, i.e. the immune system's ability to mount a more rapid and effective defence against a repeat attack by a given pathogen.[5]

Delta waves during deep slow-wave sleep induce substantial production of growth hormone as well as prolactin, a peptide hormone involved in growth and immunity.[6] Prolactin levels mirror those of growth hormone: minimum levels around noon followed by a tiny increase in the afternoon and then a surge at the onset of deep slow-wave sleep.[7] The absence of stress hormones (cortisol, adrenaline, and noradrenaline) at this time of night promotes the activation of an immunological memory process that would be impeded by the anti-inflammatory effect of cortisol, which is always active during waking hours. And lastly, the effects of sleep on the production of growth hormone and prolactin have been observed in the same proportions on a growth factor well known to paediatricians: IGF-1 (insulin-like growth factor), which specialises in stimulating the growth of cartilage and bone in children.[8]

Each of Van Cauter's and Born's studies comes to the same conclusion, which may be readily summed up in this facile formula: *lack of sleep impairs growth and accelerates ageing.* So we should think of deep slow-wave sleep the way researchers think of a grant: use it or lose it by the end of the year. And once we hit 40, we have less and less of it at our disposal.

MAN-SLEEP

Sleep is also prime time for testosterone, the so-called male sex hormone (the average testosterone levels in women are ten times lower). Like growth hormone, testosterone takes it cue from our slumbers.[9] Its production is low during the day — and yet essential, given its crucial importance for alertness, vigour, and muscle development in general — and swells when the brain is resting. In this case, deep slow-wave sleep is not the preferred stage for hormonal production: testosterone turns out to be more sensitive to sleep deprivation in the second half of the night — i.e. when light slow-wave and REM sleep are predominant — than in the first half.[10] So testosterone waxes plentiful in the early morning after a complete and restorative night's sleep, and wanes in the evening.

But back in 2005, when Swedish scientists at the National Institute of Psychosocial Medicine in Stockholm brought together seven subjects aged 22–32 and had them sleep during the day instead of at night, the subjects' testosterone levels showed the same abundance during their daytime sleep.[11] A few years later, Eve Van Cauter and her neurobiologist colleague Rachel Leproult at the University of Chicago set out to gauge the effects of sleep deprivation on testosterone production. Their ten subjects, with an average age of 24, were allowed only five hours of sleep for eight nights in a row: this time around, while their *nocturnal* production of testosterone was not too significantly affected, their *daytime* production dropped by 10–15 per cent,[12] enough to reduce an athlete's tenacity during a competitive heat, a businessman's entrepreneurial zeal during tough negotiations, or a student's staying power during an oral examination. Furthermore, if we compare 20-year-old with 40-year-old sleepers, as an Israeli team did at the HaEmek Medical Centre in Afula in 2003, we find that the deterioration of sleep quality as a function of age comes with a roughly 30 per cent drop in testosterone production during sleep.[13] It turns out that men's testosterone levels decrease by 1–2 per cent per year.[14] In everyday life, pulling an all-nighter or getting only limited or fragmented sleep for several days has an adverse effect on testosterone production.[15] In the medium term, poor-quality sleep on a regular basis can adversely affect the development of muscle mass, vigour in general and libido in particular, and, once again, fertility.[16]

SLEEP DEBT AND OBESITY

Sleep is like a glassy sea, immobile on the surface and seemingly suspended in time, but bubbling with life and multiple currents underneath. This doesn't mean that, while the production of certain hormones is going full throttle, some parts of the body aren't idly resting. A case in point is the digestive system, for which sleep is a fasting period, an indispensable timeout, a beneficial break. But this respite requires some outside support, which is where yet another hormone comes in: leptin. Its job is to send a message of satiety to the brain over the course of each meal, but, during our slumbers, especially deep slow-wave sleep, it suppresses the appetite.[7] And for good reason: at the end of the day, our exhausted digestive system becomes more intolerant of glucose (one of the three essential nutrients, along with amino acids and lipids) because insulin, the hormone in charge of regulating our blood-sugar levels, suddenly loses its effectiveness.

Our appetite is stimulated by leptin's rival, ghrelin, a hormone released before each of our three daily meals, but inhibited by sleep at night.[8] And yet sometimes we feel the urge to indulge in a sweet treat late at night. Have you never had a late-night snack when, for professional or family reasons, you had to stave off sleep? Night workers are the most familiar with this sort of thing: what could be simpler than to gobble up some sweets for a rapid burst of energy. Sugar is also an ever-handy ersatz consolation in case of passing or chronic melancholy, which so often causes insomnia. All this explains why, in many cases, sleep debt leads to diabetes, obesity, and, in the long term, cardiovascular disease. It is all too often attended by after-hours snacking when we shouldn't be eating at all.

Leptin is there to help us resist this particular temptation. Heeding the call of our biological clock, it shows up at nightfall, and is even more active when we're out of action.[19] Unfortunately, not only is prolonged sleep deprivation conducive to nocturnal noshing, but it also tends to disrupt our diurnal production of leptin and ghrelin, increasing our appetite and hunger all day long.[20]

WHEN THE CLOCK CHIMES IN

Other hormones, though sleep-linked as well, don't see a drop in production in the absence of sleep. This is true of melatonin and cortisol, which, as we've seen, prepare us for sleep and wakefulness, respectively. Whether you get a full night's sleep or are kept awake for 24 hours, the cortisol release is virtually identical, because its production is controlled by our biological clock, and not sleep. The same goes for thyroid-stimulating hormone (TSH), which, as its name suggests, stimulates hormonal production by the thyroid gland to boost cell metabolism. During the day, its concentration is stable and pretty low. In the late afternoon, when our body temperature peaks, our biological clock summons up this energising hormone until early evening.[21] Fortunately, sleep provides a respite for the body by stopping TSH secretion during the very first seconds of deep slow-wave sleep.[22]

Our nightly detox

While people often say they need to sleep to 'get their strength back up', the body doesn't wait till nightfall to replenish its stock of energy. When it needs some energy, it produces it, whether during the day or at night, and immediately consumes it. But sleep allows us to expend less energy (32 per cent less in eight hours of sleep)[23] and, most importantly, to devote our energy to something that can't happen without it: repairing the damage caused by our waking hours, and cleansing the body, from tip to toe, of miscellaneous waste that has built up after the day's intense physical and mental activity.

A WELLSPRING OF ANTIOXIDANTS

In order to survive, besides eliminating waste from the digestive system, our bodies must also get rid of various metabolic wastes, in particular urea (a nitrogenous waste product resulting from the breakdown of proteins by the liver and, if present in excess, liable to contribute to the development of metabolic syndrome) and free radicals derived from oxygen (by far the most toxic waste products, causing oxidative stress).[24] By damaging vital molecules in the body, oxidative stress fosters the development of chronic diseases such as cancer, coronary heart disease, diabetes, and kidney failure, among others. As we shall see in Chapter 7, taking a nap can provide therapeutic benefits by helping our bodies cope with oxidative stress.

While urea is filtered by the kidneys before being eliminated in urine, the only way to protect ourselves against free radicals is to transform them. And the body needs a whole battalion of antioxidants to do the job: enzymes like glutathione, which can be made by most cells, and plasma compounds like taurine and vitamin E. Here again, the operation isn't confined to our nightly slumbers, but many studies have shown that the concentration of these 'oxygen free-radical traps' in the blood peaks during sleep.[25] Oxidative stress in sleep-deprived rats has also been observed to increase by 150 per cent the amount of malondialdehyde, one of the main free radicals resulting from lipid peroxidation, in the liver,[26] whose antioxidant defences aren't adequately reinforced to cope with the build-up and don't return to a state of equilibrium until the next period of restorative sleep.[27] Human urine contains nearly twice as much malondialdehyde during the day (peaking at the end of the day) as at night.[28]

As for urea, a study in mice has shown that sleep deprivation increases urea levels in the blood, which indicates impairment of its elimination by the kidneys.[29] Other waste products present in the interstitial fluid — including those due to chemicals in the environment to which we are daily exposed (pesticides, hydrocarbons, heavy metals, phthalates, bisphenol A, and thousands of other compounds that have yet to be adequately assessed)[30] — are transported in the blood and lymph, day and night, to the liver, the body's detoxification centre, before being filtered, in the best-case scenario, by the kidneys and eliminated in the urine.

THE WATERS OF THE BRAIN

The brain produces an equally impressive quantity of toxic substances. Bear in mind that the brain consumes 20–25 per cent of the body's energy during our waking hours, even though it weighs, on average, only 1.5 kilograms. This isn't surprising when you consider its 100 billion neurons, which, activated by the mere blinking of an eye, a shiver, an itch, or any of billions of other daily stimuli, proceed to transmit electrical signals via their axon terminals. At the end of the day, the cerebrospinal fluid — i.e. the blood in the brain, which is highly filtered to remove any contaminants — is nevertheless full of potentially toxic residues leaked from these synaptic exchanges. Unfortunately, that includes beta-amyloid peptides: they accumulate on our neurons, fostering inflammatory reactions and the production of neurotoxic free radicals. So it's vital to get rid of them regularly, just as we brush our teeth regularly so as not to give plaque enough time to form.

The relaxation of brain activity during sleep (through the slowing of our brainwaves) does not suffice in and of itself to explain why our brains are refreshed after a long or short sleep. The accelerated cleansing of the brain during sleep has a lot to do with that, too. During the cleansing process, the brain pushes some waste towards its interstitial fluid, which then carries the waste away. What's most original about this scenario — as discovered in mice in 2013 by a team at New York University — is that the space between our billions of brain cells increases by 60 per cent during sleep.[31] Not that we get a swelled head as a result! On the contrary, we simply become ourselves again in the Land of Nod. During the day, when operating at full capacity, the brain is contracted, curled in upon itself; during our sleep, it relaxes, dilating to its full volume, which enables the cerebrospinal fluid to drain toxic metabolites away from the brain into the interstitial fluid twice as fast.

Over the long term, sleep-deprived nights compromise the restorative and protective effects of sleep. Sooner or later, they allow an excess of waste to pile up in the brain, which may not only destroy neurons and impair memory, but also form extracellular deposits that are the root cause of Alzheimer's disease. Then again, these outcomes can be staved off by napping.

Memory consolidation

So let's talk about your memory. You consciously make demands on it only in critical situations, just as you hurriedly inflate your lungs before sticking your head under water. Before taking an exam or making an important phone call, you try to remember the information you're going to need. The rest of the time, it's automatic. Except that your memory is a little more capricious than, say, breathing, and you sometimes forget what you wish you could remember. This is owing to neuroplasticity and the vagaries of what the brain picks to replay in your sleep in order to turn short-term into long-term memories.

FORGET ABOUT IT

Every moment you live through and every thought you think, everything you learn or experience, and every move you make is encoded in the brain in the form of synaptic connections between neurons. In other words, as far the brain is concerned, they are nothing but a succession of nerve impulses, which in this case constitute the language of the brain. All this brain activity over the course of the day requires a sizeable amount of energy, as I noted above, but also a certain plasticity of the brain, which, in order to manage the infinite variety of messages we receive, and learn certain mental operations while it's at it, has to create, destroy, and constantly reorganise neural networks and their connections. The movements I've just made in typing the words of this sentence, for example, are among the countless encodings whose retention would be superfluous. On the other hand, my knowledge of, say, the location of each letter on my keyboard is one of the things I've learned that has been — and continues to be — consolidated. This selective sifting of what to retain appears to be particularly effective during sleep, which narrows the synaptic interface between neurons.[32]

MAKING ROOM FOR MEMORY

When awake, your brain has to provide for the energy needs of innumerable operations, and the intensity of the connections between neurons conducting a given learning process is relatively low — which is why the underlying information is so fragile and our immediate memory so unstable. This is also why, if you're trying to memorise a passage of text or music, you have to keep going over it again and again, as if your memory were a little hard of hearing: only repeated operations are accorded a high-intensity connection. One of the great benefits of sleep is that it replays some of the connections made during the day: most of them involve inconsequential or even parasitic information and are therefore replayed at a much lower synaptic strength, leading to their gradual disappearance and consignment to oblivion; a smaller number of connections involve important information and are replayed at much higher intensity than during wakefulness. In other words, a majority of our neural connections are sacrificed in order to save energy with which to consolidate a minority. For the brain also needs to rest, and sleep serves two conflicting imperatives: it must lower overall energy consumption by the brain even as it consolidates certain information we need to retain.

REPLAY, OR: THE ART OF REVISING IN YOUR SLEEP

This process is the basis of replay theory, which grew up out of a rat study conducted in Tucson in 1994. Researchers at the University of Arizona placed a rat, whose brain was connected to electrodes, in a position to find any of several tracks leading from an intersection to a food reward. After recording the sequence of neural firing involved in solving the problem — i.e. the activity of the neurons known as 'place cells', the brain's own GPS — the scientists found the same neurons being fired during the rat's deep slow-wave sleep: the firing sequence replayed in a sort of dry run, without making the rat get up and scamper off towards its reward.[33]

Seeing as this type of experiment involves performing a

craniotomy, the surgical removal of part of the skull to get at the brain, it's impossible to verify the findings in human beings. Functional magnetic-resonance imaging (fMRI), on the other hand, is a non-invasive approximate substitute that allows us to observe blood flow in areas of the brain: since blood supplies the brain with the oxygen and nutrients it needs to function properly, this is a good, albeit indirect, way of measuring neural activity.

A Belgian study conducted in 2004 by neuropsychologist Philippe Peigneux and his team at the Free University of Brussels observed a similar reaction in the brains of 36 carefully selected human volunteers. The idea was to have them carry out a virtual spatial navigation exercise on a screen during the day and then observe their sleep at night. The subjects had to find their way through the maze of a virtual town, after which the researchers scanned their brains to determine whether the same areas activated during the exercise also fired in their sleep. It's not surprising that the subjects had to be hand-picked: who of us could bear to be cooped up in an awfully noisy chamber for the duration of a virtual exploration (which entailed learning a route to a specific location in 90 seconds or less) and then spend all night there without moving, eventually falling into a deep sleep for six hours? Thanks to these remarkably cooperative and adaptable sleepers, the researchers were able to observe that exactly the same parts of the brain that were fired during virtual navigation were reactivated during deep slow-wave sleep. What's more, the results showed that the more intense the nocturnal replay session was, the better the subject's performance the next day.[34]

TELL ME WHAT YOU REMEMBER, I'LL TELL YOU HOW YOU SLEEP

As deep slow-wave sleep diminishes with age, it makes sense that memory should likewise gradually decline in most adults over 45 years of age. Deep slow-wave sleep is necessary not only for restorative sleep, but also to consolidate memory. So, conversely, a lack of sleep is likely to make us more forgetful, whatever our age. Fortunately, this only concerns part of our memory: several studies have shown that REM sleep also contributes to the consolidation of memory, though not of the same type of information.

There are two types of memory: *declarative* (or *explicit*) memory and *non-declarative* (or *procedural*) memory. Declarative memory concerns facts, ideas, and events, with two components: semantic memory (context-free factual information) and episodic memory (personal recollections of the past). Non-declarative memory is unconscious. It concerns performance, motor and perceptual capacities, habits, and emotional learning (e.g. reading, greeting someone, cycling, playing an instrument).[35] Whether it's a matter of performing a sequence of finger movements on a keyboard or, say, memorising a list of word pairs, sleep consolidates learning in different ways.[36]

Psychologist Werner Plihal and neurobiologist Jan Born showed in 1997 that declarative memory is consolidated by deep slow-wave sleep and procedural memory is consolidated by REM sleep. All they had to do was have their subjects learn a list of words and then let one group sleep the first half of the night (which, as we have seen, is rich in deep slow-wave sleep) and another group the second half (rich in

REM sleep): when they woke up, the first group's memory performance was twice as high as that of both the second group and the control group (who were kept awake all night). On the other hand, when tasked with redrawing a geometric shape they'd learned the day before, the late-night sleepers performed twice as well upon waking as the early sleepers.[37]

SLEEP SCHOOL

Thus, our declarative memory is impaired by a lack of deep slow-wave sleep — in other words, as we get older or accumulate sleep debt. And yet nature proves indulgent: one study has shown that REM sleep can also boost declarative memory — provided it's chockful of emotional content.[38] This may help us understand why we're particularly partial to teachers who are capable of moving us, and why their courses tend to be better understood and retained than those of their lacklustre colleagues.

But what about emotional memories that are *too* intense, painful, traumatic? Should we worry that REM sleep will stir up such memories, which we'd be better off forgetting for good? Here again, nature has thought of everything: cortisol — which, as we have seen, is most plentiful in the latter half of the night — serves to protect us from upsetting memories. In fact, Jan Born's team discovered that chemically reducing cortisol production (by means of a nasally delivered protein) increases retention of highly emotional information.[39]

So sleep not only cleanses our brains of inconsequential memories, it also relieves us of traumatic memories. It reserves deep slow-wave sleep for the consolidation of declarative memories and REM sleep for the consolidation of procedural and emotional memories. But we shouldn't be satisfied with merely one or the other, consoling ourselves with a glass half full: 'I may not have any more declarative memory, but at least I've still got some procedural and emotional memory left.' That would be too simple. According to a study by Christian Benedict, a researcher in the neuroscience department at Uppsala University in Sweden, the frequency of alternation between these two types of sleep also seems to optimise memory consolidation.[40] Therefore, it's important to set aside ample time for your nightly slumbers, because it's easier to alternate between these two stages during a longer sleep, which in turn will boost your memory.

MISLEADING MEMORIES

Poor sleepers are always tempted to excuse their forgetfulness by claiming to be sleep-deprived. But unless they make a regular practice of napping, they may have a hard time regaining the trust forfeited by such an admission. Unfortunately, as neuroscientist Susanne Diekelmann at the University of Tübingen in Germany demonstrated in 2008, losing sleep is conducive to producing false memories. Her study involved 145 subjects aged 18–35 engaged in four different learning experiences. They all started by learning 18 different lists of 15 words each, in which the semantic connection between the words was not made explicit. For example, one list included the words 'night', 'dark', 'coal', etc., but not the theme word that ties them all together: 'black'. The participants were then subjected to different sleep-deprivation regimes for a night or two. Then, nine, 33, or 45 hours after learning the word lists, they were shown the (previously withheld) theme words and asked whether they were on the original word lists they'd learned. In each case, the lack of sleep caused the production of false memories: some claimed, for example, to have seen the word 'black' on the list. On the other hand, a night's sleep after post-learning sleep deprivation actually sufficed to rectify false-memory production.[41] Incidentally, these discoveries added grist to the mill of the controversial ideas of American psychologist Elizabeth Loftus, who since the 1970s has continually called attention to the risks of miscarriage of justice due to testimony based entirely on memory.

Dreams, a sanctuary of emotion

Our dreams also have a singular relationship to reality. Although they fill our nights, we're not always fully aware of them or what purpose they serve. To this day, scientists remain divided over their function, though scientific advances since the turn of the millennium have given rise to some interesting hypotheses.

POST-TRAUMATIC NIGHTMARES

Functional magnetic-resonance imaging (fMRI), which has shed light on the phenomenon of memory consolidation during REM sleep, has also revealed that the regions controlling rational thinking in the brain are relatively inactive during dreams, unlike four other specific parts of the brain: the regions involved in complex visual perception, movement, emotions, and autobiographical memory.[42] Nowadays, we can get a rough idea of the content of a sleeper's dream even before waking them up to ask them about it. But for neurobiologists who observe the brain activity of sleepers and then interview them, the correlation between lived experience and dream content turns out to be far less pronounced than expected.

In 2001, a team at Harvard Medical School analysed nearly 300 dreams and found that a mere 1–2 per cent of dreams re-enact real-life events, whereas 35–55 per cent of dreams are inspired by difficult, painful, even traumatic emotional shocks.[43] On the basis of these figures, American neurobiologist Matthew P. Walker recently hypothesised that REM-sleep dreaming at night serves to relieve the stress caused by negative emotions experienced during the day.

This curative effect can be explained by the surprising fact that production of one of the principal mediators of the stress system, noradrenaline, is interrupted during REM dreams. While cortisol levels increase at the end of the night — which, as we have seen, helps us forget heavily emotionally charged information during this sleep period — noradrenaline, a neurotransmitter known to foster feelings of anxiety, seems to be suppressed by dreaming. Consequently, our emotional memories are reactivated without the attendant anxiety and, in a sense, are diluted in our dreams. If this 'soothing balm' didn't exist, says Walker, we'd spend our days in a state of permanent anxiety, because each painful memory would be accompanied by a heavy load of emotional stress.[44]

Why, then, do some memories return to haunt the nights and days of victims of violent emotional shocks? The answer lies at the intersection between Walker's work and that of American psychiatrist Murray Raskind, a specialist in the phenomena of post-traumatic stress disorder (PTSD) among US veterans. In the early 2000s, Raskind successfully treated his patients with prazosin, a drug originally designed to lower blood pressure, but which, in his patients' case, put a stop to their nightmares, thanks to one particular effect of this anti-hypertensive drug: it prevents noradrenaline from acting on the brain (which corroborates Walker's theory).[45] The veterans were previously unable to escape from the downward spiral of their emotional memory because the concentration of this neurotransmitter in their brains was too high to allow their dreams to 'dilute' their trauma.[46] Walker's very promising research could pave the way for new treatments not only for PTSD, but for some forms of depression, too. (I might note in passing that napping also reduces the concentration of noradrenaline in the brain, but I'll get back to this important point later.)

EMPATHIC DREAMING

Another function of REM sleep follows naturally from its special connection to our emotional memory: namely, our ability to read emotions on the faces of others. Surprising as it may seem, this capacity is not innate, nor is it learned once and for all, but is continually maintained by sleep. A team at the department of psychology and the Helen Wills Neuroscience Institute at the University of California,

Berkeley found out in 2010 that a single sleepless night is enough to impact our affective empathy, i.e. our ability to recognise, at the very least, sadness, joy, or anger on a person's face.[47] So imagine the damage caused by sleep debt to the minds of those whose profession (judges, police officers, doctors, nurses, teachers, etc.) requires an ability to empathise. A year later, the same team showed that REM sleep is crucial to this regular upkeep of our ability to interpret outward signs of affect.[48] Given that dreams — which, as we have seen, regulate our emotional memory — loom large during this stage of sleep, Walker posits that dreaming is the very function of REM sleep. This could be more formally demonstrated by an analysis of dream content. In any case, it's important to protect our REM sleep, and hence our ability to dream, if we're to make the right calls in interacting with others — in our domestic, social, and professional lives.[49]

DREAMING ABOUT THE WORST FOR THE BEST

Surprisingly, dreaming also helps us manage future emotions. In the run-up to an important event in our lives, we're often anxious about the outcome. Our dreams then seize on our emotions to imagine various scenarios of potential failure so that we'll be better prepared — through 'hyper-associations' of images, thoughts, and recent emotions — to face the real-life situation when the time comes. In an amusing survey conducted in 2014 at Pierre and Marie Curie University by the sleep-disorders unit of Pitié-Salpêtrière University Hospital in Paris, 60.4 per cent of the 2,324 students registered for the entrance exam to medical school had dreams about the exam beforehand. Of that group of dreamers, 78 per cent imagined failing because they'd fallen ill, overslept, broken their glasses, forgotten what they'd learned, couldn't obtain all the exam questions, came in last, and so on. All this would be of little interest if success on the examination hadn't turned out to correlate with this propensity to anticipate the event — even the worst-case scenarios — in dreams.[50]

And if you never recall your dreams, you needn't worry that you're being deprived of dreams and the 'soothing balm' your slumbers should provide. Some people recall their dreams simply because they wake up at night more often, and for longer intervals, than others. This doesn't help with restorative sleep, but it can have certain benefits, especially during an outdoor nap, because it reflects a heightened receptivity during sleep to environmental factors, including potential hazards.[51] People who nap more for pleasure than out of necessity tend to wake up more often during their naps and probably also derive a psychological benefit from the capacity to recall their dreams.

In conclusion, however, let's get back to the notion that REM sleep is the sole domain of dreaming. It was long held to be the only physiological marker of dreaming, but a number of experiments have since shown that dreams roam more freely than previously thought. One such experiment, designed in 2004 by the department of psychophysiology at Kohnodai Hospital in Ichikawa, Japan, used naps to prove the point: 11 volunteers (monitored by polysomnography) were asked to take a series of 20-minute naps, separated by 40-minute periods of wakefulness, over the course of 72 hours, i.e. three days in a row, during which each was asked about the content of their dreams, if any. While over half the naps that included REM sleep featured a dream experience, roughly 20 per cent of those without any REM sleep actually did, too. That said, of the 735 reported dreams, those from REM naps were longer and more vivid.[32] Which is, indeed, the hallmark of REM-sleep dreams: their extravagance and emotional intensity, the prominence of visual elements, and the complexity of the situations they involve. This is why vivid dreams are more likely to occur at the end of the night (which is rich in REM sleep) and to leave lasting impressions, including recollections, on our wakeful days.

Part II

A World of Sleep Debt

3

The Sandman's Debtors

There's something dry and barbed about the word 'debt'. It smacks of no relief, no reprieve, and something you'd better not rub the wrong way, like a cactus in the desert. It's all too readily associated with the economy and those who can't meet their financial obligations, as in 'bad debt'. But we owe nature a great deal more — in fact, just about everything we've got, which we might call a 'good debt'. And what do we owe ourselves? A psychologist will insist you give the matter plenty of thought, and you'll find you owe yourself much more than you think, though you're reluctant to admit it: another sort of 'good debt'. A sleep specialist will be less lenient and more likely to alarm you, seeing as night after night more and more people sink an hour or two deeper, if not more, into sleep debt: 'bad debt'. An hour or two may seem negligible to you, but if you knew the true value of your precious forty winks, you'd definitely be less spendthrift with them.

At any rate, a borrower must repay their debt sooner or later. You've no choice in the matter. So just do the maths: an hour less sleep each night and by the end of a single week you've already lost what adds up to a whole night's sleep. At that rate, you'll lose more than a month of sleep each year — which can only take a toll on your health.

Another day older and deeper in debt

To find out how sleep patterns evolve in a given population, we need to observe them over time. And to find out whether those findings are universally valid, we need to gather data in as many different countries as possible. Fortunately, sleep is a matter that matters a lot to people in every country of the world. Long-term studies are rare, to be sure, but some tenacious scientists have stayed the course and really gone the distance.

Like those at the National Sleep Foundation (NSF), for instance, a world-renowned American non-profit organisation founded in 1990, whose mission is to improve the health and wellbeing of Americans by educating them about the virtues of sleep. Five studies by the NSF between 1998 and 2009, based on telephone surveys of a representative cross-section of the population, comprising 1,000 Americans over 18 years of age, revealed a spectacular increase during that ten-year period in the number of Americans sleeping less than six hours on weeknights, from 12 per cent of the population in 1998 to 20 per cent in 2009.[1] Other studies have reached similar conclusions, including one conducted in 2010 by a team of researchers at the University of Chicago's department of medicine, who analysed the data collected in sleep diaries[2] over the course of eight separate experiments: sleep time did indeed decrease between 1975 and 2006, though only among full-time workers.[3]

Here in France, the National Institute of Sleep and Wakefulness (INSV), an association founded in 2000 on the initiative of the French Society for Research and Sleep

Medicine,[4] also undertakes to make sleep a public-health issue. Its mission is to 'raise awareness, inform, and educate people about sleep–wake disorders'. In 2012, the INSV found that the sleep debt trend in France was getting even worse: 31 per cent of the population aged 18–65 were now sleeping an average of five hours and 33 minutes on weeknights.[5] Since 2004, the Institute has been varying the focus of its surveys each year to cover 'Teenage Sleep' (2005), 'Drowsiness and Work' (2006), 'Sleep and Transport' (2014), 'Sleep and Nutrition' (2015), 'Sleep and New Technology' (2016), 'Sleeping Alone or Not' (2017), etc. Regardless of the target population, each survey furnishes new data on sleep patterns and their development over time. I note in passing that, at present, the percentage of adults who sleep less than six hours a night is significantly higher in France (31 per cent) than in the United States (20 per cent).

A DEBT FOR LIFE

Our sleep requirements change with age, so researchers naturally factor age into their equations. In 2010, the Hôtel-Dieu Sleep Centre surveyed 9,251 French 11-to-15-year-olds for an international study.[6] They found a regrettable tendency among adolescents to fall deeper and deeper into sleep debt: sleep duration drops by more than 90 minutes a night between 11 and 15 years of age, which is incompatible with their physiological needs during this stage of development. Furthermore, the survey found that, as they grow older, kids sleep less on weekdays: from nine hours and 26 minutes on average at the age of 11, to seven hours and 55 minutes at 15.

As for young people aged 15–24, the 2018 INSV survey found that while 88 per cent of them thought they weren't getting enough sleep (on weekdays or weekends), 99 per cent of them were actually feeling the effects: inattentiveness, irritability, drowsiness, and dejection. During the week, 20 per cent of them slept a mere five hours on average, and at least that many complained of losing almost two hours of sleep a night.[7]

Needless to say, things don't get any better with age. One survey conducted by Professor Damien Léger's team at Paris Descartes University warned about the state of affairs in France: in 2011, nearly 20 per cent of adults aged 25–45 were sleeping less than six hours on weeknights.[8] The purpose of this study, however, was not merely to count these short sleepers, but to identify their actual sleep requirements: it turned out that 39 per cent of them (7 per cent of the total population) were happy short sleepers, who didn't feel drained or lethargic during the day, while 16 per cent were actually insomniacs (who were eventually able to change categories after appropriate treatment); but nearly half (45 per cent) were amassing significant sleep debt, for they reported a daily deficit of over 90 minutes, which was not ascribable to a sleep disorder, but to a lifestyle incompatible with their bodies' needs. Then again, certain medium and even long sleepers were also suffering from sleep debt: it emerged that, whatever the category of sleeper concerned, 20 per cent of the French adult population were taking on over 90 minutes of sleep debt a night, which extrapolates to over three million men and women aged 25–45 losing over a month and a half of their personal sleep capital each year.

A GLOBAL PHENOMENON

Our sleep began its long decline when we started looking at clocks, disconnecting from nature, and letting the economy govern our society. As the American historian Lewis Mumford wrote way back in the 1930s: 'The clock, not the steam engine, is the key machine of the modern industrial age.'[9] And as philosopher Thierry Paquot recalls in his lovely book *L'art de la sieste*: 'Over the course of the 13th and 14th centuries, the mechanical clock ... with a bell that rings every hour, supplanted the clepsydra in the foremost cities of Christendom.'[10] Did it provide information? Yes, but information resembling an injunction, like a new way of beating a drum. Then, writes Mumford, came 'wicks, chimneys, lamps, gaslights, electric lamps, so as to use all the hours belonging to the day'.[11] Now the Industrial Revolution could start. All over the world, time became money, and sleep a liability.

In 2013, the NSF polled 250 active adults in six different countries, asking how many hours they slept and how many they ought to be sleeping to be fit for work. The responses were relatively uniform: 69 per cent of respondents in the US and Mexico were sleep-deprived, as were 58 per cent in Canada and the UK, 70 per cent in Japan, and 63 per cent in Germany.[12]

It's the same story in all capitalist countries. And every (or almost every) country has its own alarming epidemiological studies to prove it. Over in South Korea, for example, a survey undertaken in 2008 by Chosun University Medical School in Gwangju revealed that 37 per cent of Koreans averaged six hours of sleep or less.[13] Up in Sweden, a large-scale survey of 38-year-old women carried out by various medical departments at the University of Gothenburg revealed a similarly worsening situation: the women had lost an average of 24 minutes of sleep per day between 1968 and 2004, and only 3.5 per cent of them were still sleeping nine hours a night in 2004, down from 11.3 per cent in 1968. Not only that, but whereas 17.7 per cent had complained of sleep disorders back in 1968, that figure had shot up to 31.7 per cent by 2004.[14] Right next door, in Finland, various university research centres have established that between 1981 and 1990, 20 per cent of the population were suffering from over an hour of daily sleep debt.[15] And, unsurprisingly, sleep debts come to as much as 95 and even 120 minutes a night among certain at-risk populations, such as Spanish students[16] and Italian truck drivers,[17] respectively.

Sleep debt is a global phenomenon, and sleep a collateral victim of globalisation. We know how the Earth looks from the sky, how little wild beauty remains and what massive scars we've left on the world's wilderness. We know a lot less about how sleep looks from the sky. However, an aerial photograph of the planet at night will show other-worldly golden haloes shining around big metropolitan areas in particular, which makes one despair over what sleep looks like there, in our teeming hives of nonstop activity, our cities that never sleep. Maurice Ohayon, a professor of psychiatry and behavioural sciences and director of Stanford Sleep Epidemiology Research Center, asked himself this question. Based on satellite images from the US military, he showed that the brightest areas in North America at night correspond exactly to the areas in which people get the least and worst sleep.[18]

Stolen slumbers: shared suffering

In other words, the fewer stars you see in the sky, the more active a city is ... and the more sleep-deprived its denizens. The reason is often stress at work and the frenetic pace imposed by the city on their everyday lives. The NSF's aforementioned study from 1998 to 2009 found that the average American's alarm clock went off at 6.05 am on weekdays, after six hours and 40 minutes of sleep. These figures roughly match sleep statistics for the French, a majority of whom likewise rise at 6.00 am to go to work and sleep an average of six hours and 55 minutes a night.[19] Many of us try to make amends by getting up an hour later on our first day off. Unfortunately, the late hour at which we turn in on those nights automatically neutralises the effects of sleeping in.

SLEEPY, SLUMPING SCHOOLCHILDREN

As I've said, the primary victims of this system that deprives people of sleep, however badly they may need it, are kids, whose daily schedules are all too often in total contradiction with their sleep needs and — in the case of adolescents — with the difficulties they have getting to bed early. According to psychophysiologist Hubert Montagner: 'School time needs to be reorganised, seeing as the timetables of school and society are not adapted to children's rhythms.'[20] Even in preschool, kids are already displaying a lack of alertness between 8.30 and 9.30 in the morning, and 68 per cent of Year 2 pupils (6–7 years old) are yawning or slumping at their desks between 9.00 and 9.30 am, compared to 36 per cent between

9.30 and 11.00 am.[21] French children, even preschoolers aged 3–6, have the longest school day in the world: five hours and 30 minutes of classes and 30 minutes of recreation. The school day elsewhere in Europe is generally four to five hours long.[22] Some children find it difficult, if not impossible, to remain alert and attentive and consequently to process information, understand, and learn every morning for three hours. Montagner recaps the situation thus:

> Over 80 per cent of the pupils who are underperforming or failing at school (bear in mind that many suffer daily from sleep deficit and emotional insecurity) yawn, slump onto their desks, stretch, fidget, do not answer when called on, shut their eyes, doze, even fall asleep between 2 and 4.30 pm. This figure often approaches 90 per cent at schools in urban areas whose inhabitants contend with personal, family, and social problems. To all appearances, these children wait for 4.30 as for deliverance.

In 2007, a study of 1,500 six-year-olds by the Université de Montréal department of psychology had already shown that 41 per cent of those who'd slept less than ten hours a night between the ages of three and six had verbal learning disabilities.[23]

EARLY-MORNING COMMUTE

These time pressures are often linked to those of parents who have no possibility of changing their working hours for the sake of their children's physiological needs. Not to mention the fact that daily schedules imposed on employees are generally inconsistent with their sleep needs, too, according to 60–70 per cent of workers in industrialised countries, as we have seen above. Many are required to show up to work at a time that's too early for them. And then there's the daily commute. In big cities, as in the countryside, many workers have to abridge their slumbers to allow for the time it takes to drive or take public transport to work: the French spend an average of one hour and 20 minutes a day commuting to and from work, though the daily commute exceeds two hours for 18 per cent and four hours for 6 per cent of the population.[24] About 60 per cent of us live roughly 20 kilometres from our workplace, and we set our alarm clocks accordingly. How many haggard faces, staring eyes, and closed eyelids succumb to the swaying of trains, buses, and trams in big cities, or commuter trains and coaches in surrounding towns? And how many motorists have to fight off drowsiness at the wheel in bumper-to-bumper peak-hour traffic in the morning and evening? We already know the answer: another INSV study in 2014 found that 17 per cent of motorists felt drowsy, and 9 per cent of them admitted to having nodded off at the wheel at least once in the past.[25]

HOW WORK ENCROACHES ON OUR SLEEP

Of course, commuting isn't the only aspect of the workaday world that deprives us of sleep: harsh working conditions, occupational stress, tension with higher-ups, overwork, and burnout are all factors that cause insomnia — one of the two main sleep disorders, the other one being obstructive sleep apnoea syndrome, which can also exacerbate sleep debt.[26] Christophe Dejours, a psychiatrist who decried the wave of suicides at work 20 years ago,[27] feels that the situation has only got worse over the years, giving rise to new work-related pathologies.[28] 'Various claims to the contrary notwithstanding,' he wrote in 2004, 'modern man's fatigue has not let up in the 21st century, owing in particular to changes in work-related constraints.'[29] One of the causes of occupational strain afflicts more and more workers nowadays: having to work at night or at irregular hours to meet the demands of a globalised economy, which, by its very nature, has to keep operating continuously.

THE GRAVEYARD SHIFT

A report on night work in France by the labour ministry's statistics office found a 40 per cent increase in the number of salaried employees working at night, either on a regular or irregular basis, from 1991 to 2012. Three and a half million people, or 15.4 per cent of the French workforce, worked at night in 2012.[30] While their ranks included fewer women (9.3 per cent) than men (21.5 per cent), the number of women working the graveyard shift had nonetheless doubled in 20 years, especially after 2001, when women working in industry were first allowed to work nights.[31]

Not all European countries are alike. Some, such as the United Kingdom (with the dubious distinction of nearly 25 per cent of its labour force working nights in the 1990s), Denmark, Belgium, and Italy are now seeing an overall decrease in the number of night workers. Others, such as France, Spain, and the Netherlands, are seeing a rise. The disparities between these countries are relatively small, however. Most have night work levels near 15 per cent, with some exceptions: Italy was down to around 10 per cent in 2012 after a few years' increase,[32] and the Czech Republic up to 24 per cent in 2005.[33] Elsewhere in the world, the situation is no better. In the United States, 26.6 per cent of the working population had to work nights between 2003 and 2011.[34] The percentages came to 17.5 per cent in China and 20 per cent in Senegal in 2005, 15 per cent in Chile in 2002, and so on — again with some exceptions around the world, such as Brazil, where night work accounted for only 9 per cent of salaried employment in 2002.[35] To make matters even worse, many night workers around the world are poorly paid, so they need to top up their earnings by working during the daytime, too, just to make ends meet.

According to the French labour report, those who most work at night are professional drivers (both public and private transport), soldiers, police officers, and firefighters, followed by nurses, nurses' aides, and midwives. Skilled workers in the processing industries (paper, food, chemical, pharmaceutical) come in next. But many other occupations are worrying, especially in the tertiary sector and civil service: the table below provides a rather staggering overview of the situation in France.

Sleep debt takes a heavy toll on night workers, and, if you do have to work the graveyard shift, you're better off being young enough to withstand the effects and hope for rapid recovery. Because you don't come away unscathed from a working life spent desynchronising from your circadian rhythms — which is why you find mostly 30-year-olds on the front lines in all these occupations. With a few exceptions — notably in construction and temp work — the frequency of night work decreases with age: so much so, in fact, that at the RATP (the public transport operator in Paris), as reported by Martin Thibault, a sociologist of work, 'While most senior staff work various daytime shifts, young people are often confined to rotating shifts or the night slog'.[36] Here again, the gradual loss of deep slow-wave sleep clearly accounts for the increasing difficulty workers over 40 have recovering from night work.

It's not enough to be young and single to cope with the effects of this kind of work, as I observe in the 'sleep hygiene' training I've been providing to night workers for many years now. Some of the people I deal with have a circadian clock that's flexible enough to adapt to changes in rhythm or are short sleepers by nature and of a so-called 'evening' or 'late' chronotype (i.e. night owls). The others, who have a hard time adapting to changes in rhythm, to keeping late hours and sleeping too little, soon give up this type of work, even if it means forgoing the much-coveted advantages: less contact with the higher-ups at such late hours, compensatory time off, and higher pay.[37]

However amenable they may be, workers who've lasted 10 or 20 years in the night shift have had to pay the price of such a ruinous sleep hygiene. And they've all had to develop fairly sophisticated compensatory strategies to go the distance. One such night worker, who had to travel 80 kilometres to work at night and back in the wee hours of the morning, told me that she regularly changed routes simply in order to perk up her attention, and routinely planned stops along the way for catnaps — at a village square or at the edge of a forest or in a shady park. All night owls have (and absolutely must have) their own little 'life hacks'. It's a matter of survival.

THE LIFE OF A NIGHT-SHIFT NURSE

In 2011, an American study undertaken by Vanderbilt University in Nashville, Tennessee, looked into the recovery strategies of 331 night-shift nurses during their time off (three days off after working four nights). The study found that a handful of them (2 per cent) didn't change their rhythms at all, but went on sleeping by day and staying up at night. A larger contingent (15 per cent) got roughly half their sleep during the day and half at night on their days off. The majority of the nurses surveyed, however, switched completely from daytime to night-time sleep: a quarter of the study population went about that by forgoing sleep entirely during each transition (going without sleep for over 24 hours on their first day off as well as upon returning to work), while half of the nurses spent the better part of their first and last day off in the Land of Nod. Lastly, a little over 10 per cent of the nurses surveyed systematically slept during the latter part of the night and the next morning — and they proved to be the ones best adapted to this work schedule. According to their responses, they felt less fatigued than the others, and their quality of life in general and of sleep in particular was far better, too. Their colleagues who tried to reverse their sleep–wake rhythms by depriving themselves of a night's sleep during each transition, on the other hand, proved less well adapted.[38]

SHIFTWORK: BROKEN CLOCKS AND BROKEN LIVES

Shiftwork, which often includes night shifts, involves several workers taking turns at the same position according to a set roster. In some cases, that means rotating shifts 24/7 according to a hellish 3 × 8-hour plan or a no-less-arduous 2 × 8-hour plan, both of which entail recovery periods that are at odds with a regular family life, conducive to social isolation, and, what's more, frequently cannibalised by parental and domestic duties. Shiftworkers are routinely compelled to change their daily schedules — alternating between morning (5.00 am–1.00 pm), afternoon (1.00 pm–9.00 pm) and night (9.00 pm–5.00 am) shifts — which continually desynchronises their inner clocks and disrupts their biological rhythms, even more than night work alone. Not to mention the strain on the body and the mental strain of having to stay awake and alert, which are often more taxing in shiftwork: having to remain standing up and/or in strained, unnatural positions indefinitely and to endure the vibrations of heavy machinery, carry heavy loads, or stare at a screen to make out tiny numbers and letters for hours on end while remaining attentive to light and sound signals.[39] Even if the resilience of youth can absorb much of the harshness of these strenuous working conditions, shiftworkers soon feel old before their time. The handful of profiles of night and shift workers on the website of the French occupational health and safety institute don't suffice to reflect the sheer variety of harsh working conditions to which workers the world over are subjected, but they do present some cases in point — such as a skilled worker who, after 20 years on the job, says that he's gained weight from snacking in order to 'stay the course', continues to smoke simply to 'keep alert', can't sleep the way he used to anymore, suffers from chronic fatigue and slight depression, and was diagnosed with high blood pressure at the age of 45; a divorced housekeeper raising two children aged five and 13 who starts her day at 5.30 in the morning to get to work by 8.00 after a 90-minute commute and doesn't get home till 9.00 pm; or a nurse who works 12-hour shifts, from 7.00 pm to 7.00 am, three days a week, including weekends, when she isn't required to work overtime to handle emergencies, and no longer has a family life or social life.

Elsewhere in the working world — in rail transport, for example — biological rhythms are so disorientated by irregular working hours that 'rail workers feel sleepy when it's time to eat and hungry when it's time to sleep'.[40] This personal account speaks volumes about the 'inventive' work scheduling to which they are subjected: 'I start my shift at 8.30 in the morning' — which means getting up at 6.30 am — 'and lay off at four o'clock on the dot. I stay over at the station of arrival for nine hours off, away from home ... So I start again at one' — 1.00 am — 'and finish at 7.30 in the morning. That's my day on paper. Well ... How do you cope with that?' Here again, above and beyond their impact on health and the permanent anxiety they cause, these ravaging working hours inevitably wear away at family life and social life.[41] But the resultant isolation is often professional, too: at companies like the RATP that practise 'fragmented' working hours, some staffers can no longer put faces to their co-workers' names because they simply haven't seen enough of them.[42]

Roughly 20 per cent of the French working population suffer under the yoke of night work or shiftwork year in, year out, and under strenuous conditions as various as the many occupations concerned.[43]

ROAD SLAVES

Lorry drivers are a special case, and a rather extreme case at that, in that some of them work up to 13 hours a day, six days in a row, followed by 45 hours off. After four hours of non-stop driving, they're required to take a 45-minute break.[44] Julien Brygo, a journalist who has spent years exploring the social utility of various occupations in France, heard this account from André Ribeiro, a lorry driver employed by a California-based company:

> The hardest part is the waiting, the loneliness, being stuck in your own thoughts. And the fatigue. All the drivers will tell you: we feel we're a hazard because we're pushed to the absolute limit. I drive 11 hours every day! Eleven hours sitting there on my bum! We're hopped up on caffeine, energy drinks — whatever's available, we'll try it. And we use sleeping pills to fall asleep.[45]

LIVELIHOODS AT RISK

In any case, according to a 2012 report commissioned by the French ministry of health on which I was able to collaborate, working nights or shifts results in the loss of one to two hours of sleep per 24 hours, on average. And it leads to chronic sleep debt, which, compounded by the desynchronisation of our biological clock, may be 'associated with a moderate increase in the risk of cardiovascular disease, body mass index, high blood pressure, and lipid metabolism disorders'.[46] Back in 1986, the prestigious British medical journal *The Lancet* published the results of a pioneering study that followed 504 shiftworkers at a Swedish printing works over the course of 20 years. The risk of heart attack (myocardial infarction or angina pectoris) increased as a function of shiftwork by a factor of 2.2 and 2.8 after 11–15 years and 16–20 years of shiftwork, respectively.

Working nights or shifts is particularly hard on women: because circadian deregulation, which causes changes in melatonin secretion, is a 'likely risk factor for breast cancer'; and because it 'may be associated with a moderate increase in the risk of spontaneous abortions, premature deliveries, and intrauterine growth restriction'.[47]

At the rate things are going, we may well wonder whether society will in future set any store by getting a good night's sleep and living in harmony with our circadian rhythms. In the meantime, while 80 per cent of the workforce may not (yet) be victims of night and shift work, the fact remains that they, too, are afflicted by sleep debt.

4

How Deep in Debt Are We?

A single sleepless night already incurs a certain amount of sleep debt. To be sure, a small debt can be quickly repaid. The problem starts when the debt starts piling up, when doing without some of the sleep we need becomes routine, even constant. And the world we live in produces chronic debt that carries multiple risks. The health problems we may meet with at any age, whether short- or long-term, major or minor, regular or occasional, invariably prompt the same questions: 'How did I come down with this? How did I get this way?' While entertaining all sorts of different hypotheses, we tend to overlook one potential cause of our afflictions: chronic sleep debt.

Sleep pays all debts

If you're overweight or suffer from frequent colds, high blood pressure, diabetes, recurrent headaches, or backaches, ask yourself the question that far too few of us do: what about my sleep? To make matters worse, many of the sleep-deprived try to stave off drowsiness by smoking more, eating sugary food, drinking coffee and energy drinks, sometimes taking all sorts of drugs — and it becomes a vicious circle. When they realise that, their first instinct is right on: 'I've got to stop all this.' But that's not enough: they should also ask themselves, 'What if I tried to sleep more?'

Not for nothing is sleep restorative: it is indeed our primary medicine. This is why studies that bring to light the dangers of sleep debt are the logical extension of the ones covered in Chapter 2 when describing the repair functions of sleep. Through its impact on the immune and cardiovascular systems, as well as on metabolism, chronic sleep debt can eventually give rise to serious diseases. Some major epidemiological studies have established a close connection between sleep debt and certain morbidity and mortality factors: domestic accidents, traffic accidents, anxiety and depression, excess weight and obesity, type-2 diabetes, high blood pressure, and cardiovascular disease.[1] The reasons for this connection are both biological and behavioural: biological insofar as sleep, as you may recall, is the period of secretion and regulation of many hormones that are vital to our physiological equilibrium (melatonin, cortisol, TSH, leptin, ghrelin, growth hormone); behavioural insofar as sleep deprivation is accompanied by wakefulness disorders, limited physical activity, and emotional fragility, so it's liable to cause irritability and loss of control.

HIDDEN DEBTS

Some of us might argue that we're more adaptable and can always recover by resting more at weekends or during holidays. A minority of us may be able to do that — or at least have the impression that we can. But we'd better make sure we haven't suffered long-term damage to any part of our bodies: the mere impression of full recovery is not a reliable assessment. Remember: the more strain you put on your body and the less you maintain it, the more it wears and tears. The effects of your next bouts of sleep deprivation will continue to worm their insidious way through some part of your body until the day the red lights start flashing for the first time, revealing that you've crossed the line, gone too far, and that it's time to pull yourself together. Be careful, because wear and tear on your body may render it from time to time susceptible to accidents and pathologies that, among other things, may increase the risk of a cardiovascular event (such as a heart attack or stroke), one of the two main causes of death — the other being cancer.

RED ALERT

Just because the cardiovascular warning lights haven't flashed yet, or have no tangible reason to flash, doesn't mean there's no point in sleeping better. While short and infrequent sleep restrictions have no significant long-term consequences, chronic sleep deprivation always wreaks havoc on the body. There are several telltale signs of chronic drowsiness and, consequently, of risks to your health. The warning lights flash in different colours before turning red. Let's say the first ones flicker in yellow, for they seem harmless enough. What are they? Frequent yawning, a stiff neck, difficulty concentrating, excessive dilation or contraction of the pupils, blinking more frequently: these are all signs of excessive drowsiness. If they occur only irregularly and at long intervals, it's not too serious; but if they occur every day, that's not normal at all. Next in line, orange warning lights, especially those related to mood and state of mind: increased irritability, slowly sliding down towards depression, heightened sensitivity to pain. Lack of sleep isn't always the main cause, of course, but you should still be concerned. As we shall see, another concomitant of sleep debt is an unfortunate tendency to fall ill at the slightest opportunity, which is a sign of an exhausted immune system — this turns the alert blood orange. But when the red lights start to flash, the body goes into a panic, leading to abnormal weight gain (even if you're attentive to the quality of your diet), the onset of hypertension (high blood pressure), and hyperglycaemia (high blood-sugar levels).

This is why it's so important to ask yourself regularly about your relationship to sleep, to be attentive to any signs of insufficient recovery and ... to introduce a nap into your daily schedule. But first, let's take a closer look at the serious risks attending sleep debt. And then we'll see how taking a nap serves to ward off those risks.

First and last reckonings

INCREASED DROWSINESS, DECREASED VIGILANCE

The first conspicuous symptom of sleep debt is drowsiness, which is not to be taken lightly. The lack of vigilance it entails can cause potentially fatal accidents. Get behind the wheel of a car or cross the street when you're sleep-deprived and the warning lights suddenly switch from yellow to bright red. A study by the American National Highway Traffic Safety Administration revealed that drowsy driving quadruples your risk of having an accident and is responsible for nearly a quarter of all traffic accidents (and near accidents).[2] The study also underscored that the highest-risk drivers are young people (especially young men) aged 16–29, shiftworkers (whose sleep, as we have seen, is compromised by working nights or irregular hours), and people with sleep disorders, such as obstructive sleep apnoea or narcolepsy. In the United States, roughly 56,000 car accidents, 40,000 injuries, and 1,500 deaths a year are attributable to drowsy driving;[3] from 1999 to 2008, sleepiness at the wheel accounted for 16.5 per cent of all fatal traffic accidents.[4] The corresponding figure in Europe during those years averaged about 20 per cent. In France, 732 out of 3,970 fatal accidents in the year 2011 alone occurred on straight roads, and 85 per cent of those accidents were related to drowsiness.[5]

It should be noted that the length of sleep attained the night before undertaking a long drive seems to be inversely proportional to the length of the drive. In the mid-1990s, French sleep specialists found that half of drivers slept less than usual the night before setting out on a journey by car.[6] In 2002, another study showed that drivers who'd slept less than five hours during the preceding 24 hours were nearly three times more likely to have an accident than those who'd had the benefit of a full night's sleep. It goes without saying that this risk is even greater for drivers whose insufficient sleep the night before is compounded by pre-existing sleep debt. Motorists always tend to underestimate the effects of a lack of sleep on their brains and overestimate their ability to react (correctly and in a timely fashion) to hazardous situations on the road. The experiment that brought this dangerous disconnect to light was, fortunately, confined to simulated driving.[7]

UNDERSTANDING DROWSINESS TO COMBAT IT EFFECTIVELY

To fully comprehend the risk of drowsiness, we need to know what causes it and what function it serves. You will have noticed that it comes in two forms: drowsiness before sleeping (due to the need for sleep) and after sleeping, during the time it takes to fully wake up. Drowsiness is actually an indispensable buffer between our two alternating states of consciousness, sleeping and waking. It enables us to shift more or less smoothly from one state to the other, preparing our body for the important physiological changes that accompany this transition: a decrease in muscle tone, respiration, and heart rate during sleep, and an increase upon awakening. Narcolepsy radically curtails this transition, literally plunging the narcoleptic into REM sleep, regardless of the close correlation between variations in our core body temperature and our degree of sleepiness.

Normally, alertness increases with body temperature from dawn to late afternoon, with a slight dip in alertness (and temperature) in the early afternoon. It then diminishes (along with body temperature) from the evening until the wee hours of the morning:[8] in all this, we see the hold of our biological clock and the tyranny of its circadian rhythm. But an intermediate, so-called 'ultradian', rhythm also braces the body — about every hour and a half to two hours — to enter into sleep mode.[9] It's less pronounced than our circadian rhythm, but active enough to make driving riskier after two hours on the road and so warrant a break. It is vitally important to take note of these discreet dips in consciousness — and give them their due by taking a nap. Naturally, all these spells of greater or lesser inattention are heavily exacerbated and rendered far more accident-prone by sleep debt.

ARE YOU A RISK TO YOURSELF AND OTHERS?

There's as yet no objective test you can take at home to detect the level of your sleep debt. An electroencephalogram (EEG) can only be run in a quiet setting, in the cold semi-darkness of a well-equipped lab. One focus of my research, as a matter of fact, is detecting drowsiness by means of non-invasive biological tests using saliva. The first results suggest that the level of neutrophils (a type of white blood cell, which plays a key part in fighting bacteria and in the inflammation process) found in the blood may increase with the intensity of sleep debt.[10]

There are subjective methods, which, though imperfect, can nonetheless give us a good idea of our general state of daytime sleepiness based on a brief questionnaire. The most commonly used is the Epworth Sleepiness Scale. Care to give it a try? All you have to do is rate how likely you are to fall asleep, on a scale of zero to three, while engaged in eight different activities. For example, if you don't usually fall asleep while sitting and reading, put zero; but if you start nodding off on the very first page, put three. A total score of ten or more indicates a significant reduction in alertness; a score of 16 or more indicates 'severe excessive daytime sleepiness', which calls for medical attention.

The Epworth Sleepiness Scale questionnaire can be found at https://epworthsleepinessscale.com/about-the-ess/

INCREASED SENSITIVITY TO PAIN

Another — slightly brutal — way to gauge your sleep debt is to inflict a little pain on yourself. If a little pinch in the arm makes you scream, you're suffering from considerable sleep deficit. I'm exaggerating, but there is indeed a connection between sleep debt and pain tolerance, as demonstrated by a number of (self-reporting and hence subjectively based) epidemiological studies in various sleep-restricting circumstances. In 2008, the department of psychiatry and behavioural sciences at Johns Hopkins University brought together a representative sample of close to a thousand people and found that routinely sleeping less than six hours was associated with increased pain.[11] In 2015, the NSF polled another thousand people and found that a 'perceived sleep debt' of more than 40 minutes was associated with chronic pain.[12]

Researchers have tried to corroborate these self-reported assessments in the lab by measuring tolerance to pain caused by heat or mechanical pressure. A team in the Clinical Pharmacology Unit at Clermont-Ferrand University Hospital has established that, if you're a healthy 20-to-40-

year-old adult, abridging a single night's sleep by half or three-quarters — i.e. sleeping only two or four hours if you're used to sleeping eight — is already enough to make you more sensitive to pain. They targeted body parts particularly sensitive to pain (in this case, the hand) with a thermal diode or laser, gradually raising its temperature.[13]

I myself have taken part in experiments of this kind with Serge Perrot's team at the Centre for the Study and Treatment of Pain at Hôtel-Dieu Hospital. Subjects who'd only slept two hours the night before felt pain starting at a temperature of 42.6 °C (in the morning) and 42.7 °C (in the afternoon), while those in the control group, who'd got a full night's sleep, didn't react till the temperature reached 44 °C.[14]

Torture is an inventive art by nature: other experiments have involved applying ever-greater pressure on the skin (using a device called an algometer) or a 100-watt heat ray to a finger in order to measure the participants' reaction time. Sleep debtors who slept two hours more than usual for four days in a row removed their fingers from the heat a few seconds after those who didn't get that extra sleep.[15]

The conclusion is clear: if you have chronic, especially musculoskeletal,[16] pain and you're in sleep debt, try to get more sleep, which will make your pain more tolerable. In the long run, the lack of sleep exacerbates painful sensations, which in turn make it harder to fall asleep, leading to further deterioration of your sleep and, in a truly vicious circle, further aggravating your pain.

FROM BAD MOOD TO DEPRESSION

No doubt you've already noticed how easily a bad night can make you grumpy the next day. So imagine the effect of prematurely curtailing your nightly slumbers over a long period of time. You won't be surprised to learn that there's a close connection between sleep debt and mood disorders, either — at any age. In one experiment involving teens aged 14-17, five consecutive nights of six hours and 30 minutes' sleep — which roughly corresponds to their usual sleep duration on school nights — increased their feelings of tension, anger, hostility, confusion, and, naturally, fatigue. This ill humour was confirmed by the parents, who found their offspring more irritable, more argumentative, and touchier than usual.[17] Furthermore, all child-psychology studies show that by exacerbating the emotional susceptibility that's peculiar to adolescents, sleep restriction has the added unfortunate effect of fostering risky behaviour and a penchant for drugs, alcohol, and tobacco.[18]

As with pain sensitivity, sleep debt invariably gives rise to mood disorders, and it's not uncommon for an unredeemed sleep debt to eventually lead to a depressive episode. In order to verify the universality of this tendency — which, after all, might just be a facet of Western society — two South Korean researchers at the Kyungpook National University School of Medicine in Daegu recently demonstrated that people who sleep only four or five hours a night have, on average, more symptoms of depression than those who get seven to eight hours of sleep.[19] And seven international multi-year studies involving over 25,000 people in Japan and the US have showed that short sleepers have a 30 per cent higher risk of depression.[20] QED. The trend is indeed universal.

Needless to say, not all depression is due to lack of sleep. But depression usually impairs sleep quality — most depressed patients sleep poorly[21] — and it's estimated that close to 18 per cent of the French population (23.5 per cent of women and 12 per cent of men), across all age groups, have suffered or will suffer at some point from depression.[22] Now, one of the neurotransmitters that are weakened by depression and boosted by antidepressants is serotonin, which is involved in a great many body functions — in particular, the regulation of body temperature, mood, pain control, appetite, aggressiveness, and (you guessed it) sleep. In order for you to enter the REM stage of sleep, for example, the neurons that release serotonin must reduce their activity. This is why REM sleep sets in earlier at night for depressives: as soon as the nerve endings of the brain stem are exchanging less serotonin, the brain is, in a way, primed to enter REM sleep. Hence the overall increase in the proportion of nightly REM sleep for these patients — at the expense of some of the deep slow-wave sleep they need — and the higher frequency of nocturnal awakenings.[23]

Then again, people suffering from depression could benefit from this disadvantage by sleeping for only four hours one night a week. The absence of REM sleep that night will have an antidepressant effect by augmenting serotonin function.[24] In the early stages of treatment with antidepressants, it may even accelerate the action of the medication, which works by inhibiting the reuptake of serotonin.

SITTING DUCKS FOR VIRUSES

If there's one thing that's bound to sap a person's morale, it's catching every bug that's going round and spending whole days and nights year in, year out, shivering under the blankets. If you're prone to colds and you don't get enough sleep, you should know there's a good chance that your 'fragility' isn't due to your constitution, but to your sleep deficit, chronic or not. In 2015, researchers in the department of psychology at the University of Pittsburgh monitored the sleep of 164 healthy adults aged 18–55 for a week before inoculating them with a rhinovirus and monitoring them for another five days to see whether they caught colds or not. The results were incontrovertibly clear: the subjects who'd slept less than six hours a night were four times more likely to catch that cold than those who'd slept for more than seven hours — though the likelihood was not significantly higher among those who'd slept six to seven hours a night.[25] Then again, a few years earlier, the same team, using a subjective measure of sleep (sleep diaries kept by the 'patients' for a fortnight), had come to the conclusion that people who sleep less than seven hours are three times more likely to develop a cold than those who sleep eight hours or more.[26]

In any case, long-term sleep debt clearly makes matters worse, and people working in occupations that take a toll on our sleep are the most vulnerable. One tell-tale illustration of this points is the nurses' health study conducted by Harvard University, the results of which came out in 2009. It involved a cohort of 57,000 healthy American nurses with no personal or family medical history liable to skew the data. The risk of contracting pneumonia turned out to be 50 per cent greater for nurses who slept five hours or less per night (more than 5 per cent of the study population) than for those fortunate enough to get eight hours of sleep (about 24 per cent).[27]

To understand why sleep debt undermines the immune system, researchers at the University of California, San Diego department of psychiatry kept 42 healthy young men awake from 10.00 pm to 3.00 am for just one night. When the men woke up between 7.00 and 9.00 the next morning, a simple blood test showed that their levels of killer T cells (white blood cells that target infected cells in particular) had plummeted by nearly half.[28] But after a night of restorative sleep, their immune systems had regained their balance. In case of sleep debt, however, immunity remains constantly weakened, as revealed by the blood tests of nurses who regularly alternate between day and night shifts.[29]

Furthermore, as we saw in Chapter 2, deep slow-wave sleep helps to reinforce our immunological memory by promoting the production of growth hormone: in the event of infection, the body produces — in addition to those killer T cells — white blood cells called memory B cells, which note the characteristics of pathogens in order to produce specifically adapted and consequently more effective antibodies later on, when needed. During deep slow-wave sleep, this adaptive immune response is facilitated by the high levels of growth hormone and prolactin, as well as by the virtual absence of cortisol (and, as a result, of anti-inflammatory processes).[30]

After vaccinating two groups of people against hepatitis A, Jan Born and his team at the University of Lübeck found that a single night of sleep deprivation after the injection had the effect of halving the production of antibodies against the disease four weeks later.[31] Another study on the influenza vaccine by a team at the University of Chicago's medical school yielded similar results, though this time in a state of sleep debt. Subjects who got only four hours' sleep for six nights in a row after the injection were tested ten days later: they'd produced only half as many antibodies as the control group, who'd slept their fill.[32] So it stands to reason that lab rats should give up the ghost after going without sleep for too long.[33] Deprived of sleep for research purposes, they end up dying after two or three weeks, more often than not because they were unable to fight off even the most benign infections.

THE ULTIMATE VICTIM: THE HEART

Thus, while sleep has been said to resemble death,[34] routinely depriving yourself thereof puts you on a fast track to death's door. And there's no dearth of epidemiological studies proving the point: we now know that the probability of cardiovascular problems significantly increases as a function of sleep debt.[35] Witness a large-scale Boston University Medical School study of 82,969 nurses from 1986 to 1996, over the course of which 1,084 of them died of cardiovascular disease. The study yielded some very telling statistics: the likelihood of dying from a cardiovascular event was 1.45 times greater for the nurses who slept only five hours (or less) every 24 hours than for those who slept eight. And the closer the women came to getting eight hours' sleep, the lower their cardiovascular risk.[36] In 2002, a University of Glasgow study of 6,000 workers in industry and administration over four to seven years showed a greater proportion of cardiovascular mortality among those who reported sleeping less than seven hours a night.[37]

And in 2008, the results of an even more spectacular study by the University of Chicago appeared in the prestigious *Journal of the American Medical Association*. The researchers had 495 healthy 40-year-old men and women wear actimeters (devices that measure and record activity and rest) on their wrists to keep track of their daily sleep durations for five whole years.[38] This time around, the object was to establish a link between sleep time and calcification of the arteries, which gives rise to the plaque that causes atherosclerosis and, consequently, possible heart attacks. On average, 12 per cent of the participants' arteries calcified within five years of the first measurements; the likelihood of calcification turned out to be inversely proportional to the number of hours they'd slept each night.

More generally, in a 2012 study I coordinated involving 1,046 patients consulting their general practitioners, we found that those who slept five hours or less had an 80 per cent higher risk of developing high blood pressure than the others.[39]

In 2010, researchers at the University of Warwick and the University of Naples joined forces to sift through major bibliographic databases specialising in the biological and biomedical sciences. They were looking for studies from around the world on sleep time, covering periods of at least three years, since the mid-1960s. This gigantic meta-analysis covered a total study population of 249,324 people, 4,141 of whom had been victims of cardiovascular events, whether fatal or not. They found that short sleep durations of five to six hours or less per night increase the risk of a cardiovascular event by 48 per cent.[40] Needless to say, the risks are greater for people who work nights or staggered hours, whose sleep debt is often greater.

THE SPECTRE OF OBESITY AND DIABETES

Laboratory studies have confirmed, clarified, and extended these findings. Witness a 2009 Harvard University study of the impact of flexible work schedules on workers' health, in which a group of ten healthy men and women (average age: 25) went to bed four hours later each night for seven nights in a row, producing recurring '28-hour days'. The results were telling: within a week, although they'd slept for eight hours each 'night', their leptin (satiety hormone) levels were down 17 per cent, their glucose were levels up 6 per cent — even though insulin (the hormone regulating blood sugar) was up 22 per cent, albeit in vain, showing that their systems were out of control — and their blood pressure increased slightly as well.[41] It's not hard to imagine the impact these effects have on older workers whose schedules are constantly changing, whose health is already affected by their working rhythms, and whose sleep deficit has been building up for years.

To reproduce sleep-debt conditions in miniature, University of Colorado researchers had 16 healthy young adults sleep from 2.00 to 7.00 am on five consecutive nights. This short-term sleep debt resulted in an overall increase in food intake (exceeding the amount needed for added energy to stay up late) and an average weight gain of 0.82 kilograms.[42] Other studies on healthy young people show increases of glucose levels even in the case of partial sleep restriction.[43] As a matter of fact, two nights of four hours' sleep are enough to cause glucose intolerance during the day and reduce glucose's sensitivity to insulin.[44] In the long run, these effects considerably increase the risk of type-2 diabetes, a disease that's liable to cause vascular lesions and accelerate atherosclerosis.[45]

STRESS

In 2011, with Karim Zouaoui Boudjeltia of the Laboratory of Experimental Medicine at the Free University of Brussels, I conducted an experiment similar to the Colorado one. We had ten healthy young men (18–27 years old) sleep only five hours a night for five consecutive nights, with three nights of eight hours' sleep beforehand (for control purposes) and afterwards (for recovery).[46] Our object was to find out whether even a moderate sleep restriction was sufficient to trigger oxidative and inflammatory stress in LDL lipoproteins, whose main function is to transport cholesterol to the body's cells, but which are also the principal target of oxidation. Once they're oxidised, these LDL lipoproteins can cause lipid plaque (for the most part, cholesterol) to form on the walls of our arteries, which in turn gives rise to the atherosclerosis plaque that's responsible for long-term angina pectoris, myocardial infarction, stroke, and inflammation of the arteries in the lower limbs. Oxidised LDL molecules are therefore key catalysts of cardiovascular events. Our test was conclusive: LDL levels shot up by 80 per cent after just three nights of sleep restriction. Not only that, but after the first night of recovery, the enzyme myeloperoxidase still showed peak concentrations, which are conducive to oxidative stress

— and though oxidative stress facilitates the destruction of pathogenic microbes, it can also lead to pathological situations. Worse still, LDL molecules and myeloperoxidase didn't revert to equilibrium levels until the second night of recovery, which is further proof of the importance of the recovery process in the wake of sleep restriction: a single long sleep is not enough to offset sleep debt, even a relatively small debt of 10–15 hours' sleep.

Bear in mind that *cellular stress* has nothing to do with *stress in general*. Among other things, they serve different purposes. Cellular stress kills us little by little, and we could do without it. Stress in general, although it can be harmful, is sometimes essential: we're always moaning and groaning about everyday stress, but like poison it can have certain virtues in small doses. There's what's known as 'good stress', which lends us wings — impels us to act, adapt, surpass ourselves — and can help us overcome difficulties and emergencies, fatigue and anxieties. Without this reasonable kind of stress, we simply wouldn't get out of bed in the morning. And that's where and when it sets in, actually two hours ahead of time, to help us gradually wake up. So the day always begins with this good stress, and its principal vehicle at that hour is a hormone we've already met: cortisol.

As soon as a stressful event is identified by the brain's prefrontal cortex, the hypothalamus sends a series of chemical messengers into the bloodstream that cause the adrenal glands (located, as the name 'ad-renal' suggests, above each kidney) to release cortisol. Cortisol then interacts with all the cells in the body, inducing them, according to the nature of the stimulus, to react appropriately, which may require a small expenditure or an abundance of energy. Meanwhile, the same stressful event, whether harmless or intense, will simultaneously cause the adrenal glands to release into the bloodstream two hormones of the catecholamine family, adrenaline and noradrenaline, whose main function, depending on the circumstances, is to increase not only heart and respiration rates, but also blood pressure, blood sugar, and oxygen uptake by the brain and muscles. These two 'stress systems' — driven by cortisol and catecholamines — always work in concert, and each has a special connection to sleep. Cortisol, as I've said, takes part in our sleep rhythms if they're in phase with our biological clocks, and in the next chapter we'll be looking at the function it performs all day long. Catecholamines are less closely linked to sleep, but react badly to a lack of sleep, which, as studies have shown, augments the catecholamine concentration in the blood, as well as significantly increasing blood pressure, exactly as in any stressful episode.[47] Epidemiological studies have demonstrated that overloading the stress system promotes heart problems, in both the short term — for the most acutely stressful episodes — and the long.[48] And here again, the research unequivocally shows that a single night's recovery sleep isn't enough to bring about a return to a state of equilibrium.[49]

So what's the solution to all these health risks? SLEEP! There's no other solution. And if your nights aren't long enough for all the z's you need, well then you'll simply have to make time during the day and cultivate the art of the siesta. A siesta is a wonderfully protective night in miniature that you can take wherever you go and whenever you please.

Part III

The Siesta, Or: The Force Awakens

5

Pocket Medicine

Whatever it is that your night-time slumbers don't have time to remedy or repair, it can be taken in hand by a daytime nap, provided you set aside a little time for it. The siesta is not just night sleep's sidekick, going back over its slapdash handiwork from time to time to put it right; nor is it an old acquaintance we're happy to meet up with again every now and then. It's key to keeping in good health, a safeguard that consolidates and improves the work begun by our night-time sleep. It deploys the same slow waves and the same miniature medical team that toils tirelessly by night while we slumber. Then again, napping can't replace your nightly slumbers. It plays its own part, a few bars in the ongoing symphony of life.

Above and beyond the scornful misconception in the workaday world that it's a 'waste' of time, what keeps many people from napping regularly is simply that they are unversed in the art of the snooze. Although sorely tempted, many forgo the practice because they feel it just isn't for them: they never manage to nod off fast enough or always wake up more knackered than they were before their forty winks; it takes too long or they figure there's no point in dozing for five, ten, 15, or 20 minutes when they actually need to make up for what adds up to a whole month of lost sleep.

But they're wrong: the power of the siesta lies precisely in its capacity to produce certain effects of a night-time sleep in record time. You need only master the art of the siesta, the fine points of timing your naps and wake-ups, and learn how to reap the benefits thereof in full. The more accustomed you are to napping, the easier and the more effective and profitable it will be.

Practise, take heart, relax

If you're too unfamiliar with the siesta and doubt you'll ever be able to practise daily, the first thing to do is take control and 'make it your own'. And take heart, because you'll never benefit from it if you doubt the bond between you and the snooze.

FORCE OF HABIT

Falling asleep often proves the first stumbling block. Witness this landmark study at the University of Pennsylvania in 1977: out of 430 students surveyed, 261 (60 per cent) took naps occasionally or regularly and 169 rarely or never did because they didn't feel it was of any use at all. Among these participants, 11 habitual appetitive nappers (defined as those 'who nap lightly for psychological reasons apparently unrelated to reported sleep needs'), ten replacement nappers ('who apparently nap regularly in response to temporary sleep deficits'), and 12 confirmed non-nappers were asked to take a one-hour nap. Their sleep was monitored in the lab by polysomnography — a technique which, as you may recall, records brain activity and that of the chin and eye muscles. This experiment made the force of habit all too obvious: the two groups of nappers fell asleep within 14 and 12 minutes respectively, while it took the non-nappers 26 minutes to drop off (though three of them couldn't fall asleep at all). Furthermore, the appetitive nappers experienced more alternation between light slow-wave sleep (stages one and two) and deep slow-wave sleep (stage three), with more brief awakenings during their nap as a result,[1] than the replacement nappers, and considerably more than non-nappers, whose nap sleep was particularly poor in variation.[2]

IMPOSSIBLE TO FALL ASLEEP

So the ease with which we nap depends a great deal on our habits and penchants, and not on any innate ability. It's understandable that some people should be reluctant to take a half-hour nap if it takes them about that long just to fall asleep. It goes without saying that someone who's overly stressed out will have a harder time nodding off than someone who's relaxed. Once again, however, learning to recognise moments of sleepiness and making a habit of napping in various circumstances will help you overcome the psychological state in which you happen to be. Also, until you've mastered the process of falling asleep, relaxation techniques can be invaluable in approaching and optimising sleep.[3] In a study by our research team at Paris Descartes University, nine men and two women aged 37–52 were asked take three half-hour naps in the early afternoon (1.30 pm) three days in a row, but in a particularly stressful environment — open, noisy, uncomfortable, and exposed to daylight. Some of them had had the benefit of a relaxation session beforehand conducted by a professional; the others hadn't received any coaching. Polysomnography measured various parameters, including the different stages of sleep and a nap's 'efficiency', meaning the ratio of total sleep time to the time spent in bed. The study showed that the naps preceded by relaxation exercises increased that efficiency ratio by 44 per cent, giving rise specifically to more light slow-wave sleep — a supplement that's bound to be conducive to better recovery.[4] A prior relaxation session involving ten minutes of hypnosis can actually increase the length of deep slow-wave sleep — provided the nap is long enough to produce any deep sleep.[5]

POST-NAP GROGGINESS

Another objection non-nappers raise is the groggy state they're in afterwards — known as 'sleep inertia'. This tendency to feel after a nap as though you'd been knocked out is, once again, due to a lack of practice — a bit like a novice mountaineer who climbs too far, too fast. In 2006, a study of 40-odd students at Brock University in Ontario, Canada, found that during a 20-minute catnap, regular nappers got more dynamic, light slow-wave sleep (stages one and two) than non-nappers, i.e. with alpha waves closer to those of calm wakefulness and falling asleep. As a result, their sleep was not as heavy (and yet not fragmented either), enabling them to shake off sleep inertia sooner and providing a better-quality wake-up.[6]

Get in touch with your inner napper

ANCESTRAL LEGACY

You might also be unreceptive to ultradian rhythms (the recurrent cycles within a 24-hour day) and insensible to 'sleep windows' (daytime nap opportunities) due to heredity. In 2017, a Spanish study at the University of Murcia, the first of its kind, looked into a possible genetic basis for napping behaviour. Fifty-three pairs of female twins (averaging 52 years of age) were brought together for this purpose. Continuous monitoring of their sleep–wake rhythms over a seven-day period revealed that, at least in their case, the habit of taking a siesta and their average siesta duration were, respectively, 65 per cent and 61 per cent hereditary.[7] In other words, some nappers are, by nature, likely to have more substantial drops in body temperature than others and a keener sense of the best times of the day to catch forty winks.[8]

Be that as it may, this shouldn't deter you from reaping the benefits of napping. If it seems to elude you, just keep practising; familiarise yourself with your rhythms. Once you've experienced a successful catnap followed by an unexpected boost, a surprising second wind, repeat the experience again and again. Above all, heed the call of the siesta and get used to responding with confidence. This will build a fruitful and lasting connection with the practice of napping.

YOUR FIRST NAPS

Napping isn't entirely alien to anyone, remember. When you were just an infant, between six and nine months old, you did it three times a day: in the morning and early and late in the afternoon. Over the next four months, the late afternoon nap was dropped from your daily routine, but you stuck to the other two. By the age of one, you were usually down to one nap a day, in the afternoon. Without knowing it, you were already drifting away from a daily practice that, thanks to its abundance between the fourth and 17th months of your life, had made this a special period of growth.[9] Then, between three and seven years of age — depending on your environment (family context, respect for natural rhythms, shared room or not) and, to a lesser extent, your genetic heritage — you either remained attached to your siesta or turned your back on it while consolidating your night sleep.[10] It's estimated that 50–80 per cent of three-year-olds, but only 10 per cent of five-year-olds, still take naps, which also get shorter as you grow up.[11]

As a teenager, you had a hard time falling asleep at night and, at least during the week, getting the sleep you needed for your development, so chances are you may have taken up with this little life saver again for a while. It's estimated that 30–40 per cent of adolescents take naps between the ages of 14 and 19.[12] Unfortunately, under the burden of teenage sleep debt, your naps could get so long — one hour and 13 minutes on average in France, according to an INSV study[13] — that they were liable to chip away at your night sleep and consequently fail to provide a satisfactory antidote to your daytime drowsiness.[14] Maybe that's when you started having doubts about the siesta and began putting the practice behind you.

RETURN TO FAVOUR

Though now an adult, you've nonetheless preserved the memory of this daily timeout deep down inside. Whether or not you're in sleep debt now, every day you're reminded of napping's absence when you feel a natural circadian slump befuddling your thoughts at some point between noon and 4.00 pm (depending on whether you're an early bird or a night owl). When your brain can't quite make all the connections anymore, you've reached a peak of drowsiness — so seize the opportunity: this is the ideal window for a snooze.[15]

That's what experienced nappers do. They've got the hang of it and make the most of the opportunity to doze off without delay, aided by the attendant drop in cortisol levels and body temperature. They know that the circadian clock never stops running and that the windows close up in a heartbeat. This circadian slump after lunch is all the more perceptible if the previous night's sleep was cut short. It's conditioned not by the make-up or heaviness of your midday meal, though that may be a factor, but by your biological clock.[16]

After the age of 60, napping becomes more frequent again as the quality of night sleep declines. It's estimated that nearly half the population over 60 years of age take a half-hour siesta around 1.30 pm several times a week.[17] But in contrast with the delayed sleep phase during adolescence, the elderly often suffer from 'advanced sleep-phase syndrome', which may induce them to take a nap in the early evening at the risk of detracting from the quality of their sleep that night.[18] A 2010 INSV epidemiological study entitled 'As Sleep Grows Older' notes that this resumption of napping is accompanied by a gradual change in duration: while nappers aged 50-60 often keep it short (less than 20 minutes), those over 80 are wont to snooze for an hour or more, which can make it harder for them to fall asleep at night.

SPAIN HAS NO MONOPOLY ON THE SIESTA

Our connection to napping, or at least our views of the practice, are in large measure culturally conditioned. Some countries have made it an integral — and respectable — part of their lifestyle, while others persistently hide the habit, probably because it seems to be at odds with the cult of work. But this isn't true everywhere, since China itself, which isn't known for laxness or laziness, enshrined the right to nap in its constitution back in 1954, and very few Chinese forgo a siesta after lunch. But is there any basis for the widespread notion that countries nearer the equator, under the heat of the sun, do the most napping?

Some differences can be observed, to be sure, even from childhood. The Children's Hospital of Philadelphia, Pennsylvania, in 2013 set out to gauge the effect of cultural differences on napping habits among 2,600 children aged three to six years in 13 different parts of the world. The results showed that Asian children take more naps than their Western peers, but they also go to bed later (10.26 pm on average in India as against 7.43 pm in Australia and New Zealand) and sleep two hours less at night (averaging 8.96 hours in India, 10.88 in the UK). So, despite these differences, total sleep time per 24 hours is roughly equivalent for this cohort around the world.[19]

Unfortunately, international statistics on the napping habits of older age groups are in short supply. But in 2013, six selected countries were compared by the National Sleep Foundation. It emerged that the Japanese and Americans are the world's snooze champions: 51 per cent of them had taken naps during the preceding fortnight (about four naps for the Japanese and three for the Americans, on average), followed by the British (45 per cent, roughly four naps), Germans (44 per cent, four or five naps), Mexicans (39 per cent, three or four naps), and Canadians (35 per cent, about three naps).[20] These figures are, as it happens, fairly consistent with comparative sleep debt. They go to show that napping is still the best way to offset chronic sleep debt, from the shortest average nap (Germans: about 30 minutes) to the longest (Americans: a little over 45 minutes).

Based on these meagre statistics, however, we can't conclude that there's a real gap in the culture of napping. In the hottest climes of Latin America — especially Mexico City, Montevideo (Uruguay), Santiago (Chile), and Caracas (Venezuela) — a smaller percentage of the population take naps (roughly 30 per cent), but more frequently, averaging three a week.[21] So there's no evidence that climate significantly influences napping behaviour. Instead, sleep debt seems to be the decisive factor, at least in urban and suburban environments, where working conditions can sometimes make the situation worse.

France is no exception, except that it's one of the nations that nap the least. According to a 2013 INSV survey of the French working population aged 18–65, 68 per cent never nap, 14 per cent only take one nap a week, 10 per cent take two, and 4 per cent take three naps a week.[22] This is insufficient, given our sleep debt, which, for at least a third of the adult population, comes to about two hours a night.[23] Worse still, it appears on closer scrutiny that a mere 16 per cent of the gainful employed French take a nap on weekdays (47 minutes on average) and 29 per cent at weekends (one hour on average). The fact is that in a world of acute sleep debt, napping ought to be a more-or-less daily practice. But how do we go about submitting to this daily discipline in the present day and age?

6

The Art of the Siesta

As we shall see soon enough, a nap doesn't have to be long to be useful and effective. Depending on what your goals are and how much time you have at your disposal, you can adjust the duration accordingly and benefit from both light and deep slow-wave sleep. You can also vary the timing to suit your chronotype or to allow for missed sleep on preceding nights, or even target a particular stage of sleep in order to reap its special benefits. On the other hand, it's not always easy to find a suitable resting spot.

What's the best napping position?

Although some employers encourage napping, they seldom provide their employees with comfortable facilities for the purpose. Those fortunate enough to work from home or able to go home for lunch can enjoy the luxury of a lie-down in bed or on the sofa. The rest have to settle for making do: kipping in their seats, with their backs straight or reclined, their feet on the floor or on their desk, their heads on the desk between their hands, or whatever. Anything goes. It's up to you to find out what works best for you. Weather permitting, you can also head for a park near your workplace and stretch out on a bench or a lawn.

But beyond the question of comfort, what impact will your position have on the quality of your sleep? And how will it affect the benefits you derive from your siesta, especially in terms of subsequent alertness? Two studies in particular have addressed these questions. The first is a study by the faculty of psychology at Southwest University in Chongqing, a Chinese megacity with a population of over 30 million. It compared the effects of sitting versus lying down (for a 20-minute nap) on 36 healthy habitual nappers (equal numbers of men and women) aged 18–23. Naturally, napping in either position reduced their sensations of drowsiness and fatigue, but lying down had a greater beneficial effect on their subsequent state of alertness.[1]

If you've no choice but to nap in your seat, whether at a desk or in transit (in a car, train, boat, or plane), you have three options: you can sit upright (at least until you fall asleep), recline your seat (provided it's a reclining chair with a headrest), or slump forwards with your head on the fold-out table or against the backrest in front of you. While we still lack data on the pros and cons of leaning forwards during a nap, a recent study by the Appleton Institute at CQUniversity in Adelaide, Australia, sheds some light on the ideal angle of a reclining position for snoozing purposes. After only four hours' sleep the night before, subjects had a window of four hours (the duration of a medium-haul flight), from 1.00–5.00 pm, in which to take a siesta. Some had minimally inclined seats ('upright' = 20 degrees back from the vertical), like the economy-class seats on an aeroplane; others had tilted seats ('reclined' = 40 degrees back), supposedly corresponding to 'premium comfort'; and the rest were treated to maximum inclination ('flat' = 90 degrees, i.e. lying down), as in business and first-class seats. Not surprisingly, the superior comfort of the reclined and flat seats allowed the subjects to sleep longer (one hour more, on average), with more deep slow-wave sleep (50 minutes as against 30 minutes for those in the upright seats), fewer in-between awakenings and more REM sleep (especially in the flat position).[2]

In a word, the greater the angle of recline, the better the nap — the main reason being greater relaxation of the neck muscles. In an upright position, it's impossible to fall asleep without leaning forwards or to the side, which immediately wakes you up. This is why it's essential to be firmly supported, especially when sleep enters the REM stage, which is accompanied by total loss of muscle tone. There are oodles of accessories on the market — though they tend to be slightly suffocating — to brace the neck and alleviate some of the discomfort of napping in a seated position. But sleep quality also depends on stress systems, which are sensitive to the body's position and can, in case of discomfort, activate wake systems more quickly.

The duration dilemma

Napping is to sleep what a Swiss Army knife is to a toolbox: a multifunctional concentrate. And depending on what we want from it, we'll opt for a micro-nap (less than a minute) or mini-nap (a few minutes), a short or medium-length nap (ten minutes to half an hour), a long nap (one to one-and-a-half hours) or a very long nap (two hours). Or, of course, any duration in between. For our purposes, however, I'm going to focus exclusively on naps that, at least in the sleep lab, have proven effective in countervailing the harmful effects of sleep debt: naps of about ten minutes, naps of half an hour, and naps of over an hour. And I will take the liberty of contradicting Salvador Dalí, who vaunted the merits of the micro-siesta, which he practised with a key pressed between thumb and forefinger, which was to wake him up when it fell onto a plate on the floor:[3]

> If, on the contrary, paying a deaf ear to the call of your key, you should persist a quarter of an hour more, or even just a few minutes, this would be harmful to your work, for these few minutes of laziness would have sufficed, by themselves alone, to 'enslave' you by their heaviness for the whole rest of the afternoon.[4]

This concern, often aired by novice nappers, is only justified if you wake up during deep slow-wave sleep or, again, if you haven't had enough practice.

That said, we wouldn't want to deny the uniqueness of each napper's napping: there are a thousand different ways to practise and benefit from a siesta. The main thing is that it does you some good. The scientific studies serve merely to better understand and refine the techniques. They haven't yet resolved every aspect of napping, and there's still plenty of terrain to explore. What is certain, though, as established by a great many studies, is our knowledge of the effects of nap duration on alertness and cognitive performance.

THE LIMITATIONS OF 'ULTRA-BRIEF' NAPS

A 2002 study by the Flinders University School of Psychology in Adelaide is but one of many refutations of the contention that merely initiating a sleep episode is enough to reap huge benefits. While ten-minute naps (which usually produce light stage-one and stage-two slow-wave sleep) not only improved subjective and objective alertness, performance, and vigour, but also reduced fatigue among a sample of healthy young adults, none of these effects was observed after 'ultra-brief' 30- and 90-second naps (which are too short to get beyond light stage-one slow-wave sleep).[5] Dalí's tremendous energy must have come from some other source.

It is true, on the other hand, that sleep pressure dissipates faster than it accumulates, and that a ten-minute nap can compensate for several hours' sleep deficit. So let's zoom in on the specific benefits of a 10-to-30-minute nap.[6] Bear in mind that the nap durations indicated do not include 'sleep-onset latency', i.e. the time it takes to fall asleep, so, generally speaking, we should add about ten minutes to allow for that lead time. That said, if you are indeed in sleep debt, you should fall asleep more quickly (within about five minutes), especially if the circadian timing is right. (And I'll return to the matter of timing later.)

10- OR 30-MINUTE POWER NAP?

A short nap of about ten minutes has come to be called a 'power nap'. Its main feature is that it doesn't last long enough to produce deep slow-wave sleep (SWS) or, consequently, post-nap sleep inertia. But is a power nap long enough to affect drowsiness and cognitive performance?

To find out, researchers at Flinders University this time around compared naps of five, ten, 20, and 30 minutes' duration taken at 3.00 pm after a night of only five hours' sleep.[7] The results of eight tests the participants took after napping showed that five minutes of sleep (producing one minute of light stage-one SWS and four minutes of light stage-two SWS) improved cognitive performance only marginally (i.e. on only one of the eight tests). The ten-minute nap (producing one minute of light stage-one SWS, eight minutes of light stage-two SWS, and one minute of deep SWS), on the other hand, improved not only cognitive performance, but also alertness, on all eight tests. Not only that, but the ten-minute nap had beneficial effects as early as five minutes after waking up, and they lasted for a good 100–150 minutes. The effects of the 20- and 30-minute naps (producing one minute of light stage-one SWS, 13 and 17 minutes of light stage-two SWS, and six and 12 minutes of deep stage-three SWS, respectively) didn't begin to kick in till half an hour after waking up, because of the sleep inertia caused by deep SWS, but some of those effects lasted more than 100–150 minutes.

A year earlier, a team in the department of behavioural sciences at Hiroshima University had arrived at similar results by comparing the benefits of a five-minute nap (producing only light stage-one SWS) and a nine-minute nap (producing six minutes of light stage-one SWS and three minutes of light stage-two SWS), both at 2.00 pm after a night of a little over five hours' sleep. This time around, the five-minute nap did have an effect on alertness and subjective fatigue, but only for the first 15 minutes after wake-up. The benefits of the nine-minute nap in terms of alertness, fatigue, and performance set in, once again, 15–30 minutes after wake-up.[8]

As pointed out above, a power nap has the advantage of *not* producing deep slow-wave sleep. This 'energising' nap allows for immediate cognitive performance and its progressive improvement for 65 minutes. 20- and 30-minute naps, in contrast, do produce a few minutes of deep stage-three SWS, thus causing some sleep inertia upon waking, and show no effectiveness within five minutes of wake-up. But after 155 minutes, they clearly surpass the power nap in terms of objective alertness. At that point, the power nap can only hold its own in terms of subjective drowsiness and fatigue.[9] So it may do to restore temporarily dulled physical and intellectual capacities, and may prove particularly useful in a professional context.

To sum up, the effects of a half-hour nap kick in later, but are more lasting and, after a certain period of time, more extensive than those of a ten-minute power nap. On the other hand, it has the disadvantage of requiring 40 minutes of availability — allowing for the time it takes to fall asleep — and the added difficulty of emerging from sleepiness, which is exacerbated by sleep debt inasmuch as the proportion of deep SWS will then be generally greater. Then again, the beneficial effects seldom take more than 30 minutes to set in. In fact, as we shall soon see, the half-hour nap is, above all, 'therapeutic' by dint of its effects on pain, stress, immunity, and cardiovascular risk. These are ample reasons to practise the half-hour nap regularly, if not daily, if you're suffering from sleep debt. However, if you want to snooze immediately before an important engagement, always opt for an 'energising' power nap.

A LONG SIESTA: THE LUXURY OF A COMPLETE SLEEP CYCLE

To repay your sleep debt more quickly, don't hesitate to take occasional naps of an hour or more. A 90-minute siesta, which corresponds to a complete sleep cycle, is also sure to treat you to a long stretch of deep slow-wave sleep.[10] The wake-up is harder, but the longer sleep gives your body more time to repair, rebuild, and cleanse itself. And that's no mean feat. In addition to all the therapeutic advantages of a half-hour nap, a long nap significantly improves objective alertness for at least five hours afterwards.[11] Its capacity to diminish sleep pressure makes it the ideal nap before a hard day's night on the job. Taken in the late afternoon or early evening, it enables you — after a spell of sleep inertia — to maintain a high level of performance and alertness.

To test the effects thereof, in 2006 the Sleep Medicine and Research Center in St Louis conducted a study with 68 men and women aged 18–65. The object was to compare the effects of coffee with those of a long nap (60–90 minutes) during a simulated night shift. One group napped between 7.30 and 10.00 pm on the first two of the four consecutive night shifts; the second group were given caffeine (4 mg/kg) 30 minutes before the start of each shift; the third group got both, a nap and caffeine; and the fourth (the control group) got neither.[12] Late at night and in the wee hours of the morning, the participants then took a whole battery of tests assessing their wakefulness, alertness, cognitive performance, sleepiness, etc. Combining a nap with caffeine yielded the best results. But what's particularly of note here is that the two naps continued to confer their benefits even on the third and fourth nights, and even without the aid of a stimulant.

While long naps are clearly advantageous in cases of considerable sleep debt and for night and shift workers, they should be practised in moderation if not justified by any of the above or if not permitted by your state of health. Elderly people are advised to take naps of under 20 minutes in the early afternoon because long naps are often associated with cognitive decline and deterioration in general health for their age group.[13] A 2015 study at Columbia University involving about 100 people aged 51–89 showed longer psychomotor reaction times and a decline in memory and verbal fluency as naps get longer and exceed 60 minutes.[14] This decline could be due to a dulling of circadian rhythms, entailing a deterioration of the circuits responsible for wakefulness.[15] A Chinese public-health team recently demonstrated that naps of 30 minutes or less, in contrast, have the advantage of reducing the risk of cognitive decline.[16]

Unfortunately, long naps might also indicate a deterioration in the cardiovascular health of older people. A 2014 Chinese study of 26,801 subjects with an average age of 63.6

years found that naps of over 30 minutes increased the risk of high blood pressure by 25 per cent.[17] A Japanese meta-analysis undertaken just a year later, which followed more than 150,000 people for an average of 11 years, confirmed that the risk of cardiovascular disease nearly doubled when naps exceeded 60 minutes and that this effect was more pronounced for people over 65.[18]

Lastly, those who are neither elderly nor chronically sleep-deprived, but feel an urgent need to nap for more than one hour every day should be aware that routinely taking extended naps is less of a lifestyle choice than, as observed above, a sign of a decline in health.[19] Naps are, if you will, 'epicures' of sleep: they don't need to be awfully long to bring happiness.

Morning, midday, or evening?

But duration isn't the only parameter to consider. The timing of your timeout from your surroundings is almost equally important. Back in 1981, a pioneering study by the Naval Health Research Center in San Diego provided an astonishing demonstration of this point. The experiment looked at the effect of two-hour naps taken by sailors after an extremely long period of sleep deprivation: 45 hours for one group, 53 hours for the others. You might think that after 45 hours of work, any two-hour nap would suffice to regain some energy. But strangely enough, the various performance tests they took afterwards showed that a nap between 4.00 am and 6.00 am was less effective in restoring alertness and cognitive performance than a midday nap (between noon and 2.00 pm) — eight hours later.[20] The first group suffered from sleep inertia upon waking in the early morning because their biological clocks were imperturbably stuck in sleep mode. Also, mid-afternoon naps (around 3.00 pm) have a shorter sleep-onset latency, are more restorative, and produce less inertia upon waking than evening naps (around 7.00 pm).[21]

This point was thrown into stark relief by a 1989 study conducted by the Technion (Israel Institute of Technology) in Haifa. Its focus was, yet again, on two-hour naps after 24 hours' sleep deprivation. Nonetheless, it stressed the importance of circadian nap timing, for sleep pressure (a function of the number of consecutive hours of wakefulness) is not the only determinant of nap quality: the timing has got to be right, too, in keeping with our biological clocks — in other words, with our 'sleep windows', which depend in part on our chronotype.[22] For early risers (morning chronotype), an early-afternoon nap exceeding 30 minutes will be richer in deep slow-wave sleep than in REM sleep. The earlier your night sleep ends, the greater your sleep pressure in the early afternoon and, in turn, the more deep SWS you're likely to get during a nap of over 30 minutes. Similarly, and according to the same logic, the greater your overall sleep debt, the more deep SWS this type of nap will produce. For night owls (evening chronotype), in contrast, the same nap at the same hour will be richer in REM sleep than in deep SWS: for a more-restorative sleep, they have to put their nap off till as late as possible in the afternoon, though making sure to leave an interval of at least six hours between the end of the nap and bedtime. Once again, don't forget that how you feel upon emerging from deep SWS also depends in large measure on your age and propensities.

So, what'll it be: early or mid-afternoon? It's entirely up to you: do some trial runs, test your body, take advantage of the moments when you feel the urge to doze, and, based on your age, chronotype, and needs, choose the hour and duration of your naps accordingly. Bear in mind that, while sleep windows open roughly every two hours, the most suitable are those between noon and 4.00 pm. Still, you shouldn't write off morning naps: if your nocturnal slumbers were cut off prematurely for one reason or another, a snooze sometime between 10.00 am and noon can be quite restorative — especially if you want to benefit from the abundant REM sleep it'll have to offer after not getting your fill the night before.[23] So give the matter some thought, for it may be worth your while to prioritise certain stages of sleep.

How to boost memory and creativity

While light stage-two SWS and deep stage-three SWS help to boost alertness and cognitive performance, they have the power (along with REM sleep) to boost memory, too. For there is every indication that light stage-two SWS is also involved in reinforcing declarative as well as procedural memory, which are consolidated, as you may recall, by deep SWS and REM sleep, respectively.

THE THREE SLEEP STAGES OF MEMORY CONSOLIDATION

In 2008, a team led by German psychologist Olaf Lahl of the University of Düsseldorf's Institute of Experimental Psychology carried out a study of young adults (average age: 25) and found that even a six-minute nap (producing four minutes of light stage-one SWS and two minutes of light stage-two SWS) improved their recall of a word list they'd learned an hour earlier. Recall was greater, naturally, when the nap was extended to 35 minutes (comprising 11 minutes of light stage-one SWS, 14 minutes of light stage-two SWS, and 10 minutes of deep stage-three SWS). But the point is that sleep, from the moment of its onset, has this ability to protect memories from being immediately erased by the state of wakefulness — in which our brains are constantly stimulated by our environment.[24]

Furthermore — for there's still so much to discover about the virtues of sleep — several studies hypothesise that light stage-two SWS may be particularly effective in consolidating information recall, even in a cycle that includes deep SWS and REM sleep. A 2007 Harvard Medical School study observed that during a 60- or 90-minute nap (which produced deep SWS and REM sleep, too) after a learning session, the learning areas of the right hemisphere of the brain were only activated during light stage-two SWS.[25] More recently, in 2014, researchers from various northern European countries suggested that the processes of consolidating and eliminating parasitic memories occur separately during sleep, with consolidation occurring during light SWS, and elimination during deep SWS.[26] This hypothesis, if confirmed, would corroborate the theory proposed by Olaf Lahl's team that even brief naps have this specific capacity to boost memory.

If you want to prioritise REM sleep — which, as you may recall, actively helps consolidate procedural memory — opt for morning or extended afternoon naps (of 60 or 80 minutes). Musicians, dancers, athletes, and anyone else keen on improving motor skills should try to take some morning naps or longer afternoon naps after their work sessions. A rather original Japanese study at Tokyo Medical University in 2016 involved teaching a bunch of students who'd never juggled before how to juggle three balls. After a brief introductory workshop from 10.00 to 10.30 am, one group took a 70-minute nap starting at 2.00 pm (including 10 minutes' sleep onset, 19 minutes of deep SWS, and about 12 minutes of REM sleep). That siesta turned out to be enough for them to be able to juggle a lot better later on — during a test session at 5.30 pm, two hours after the nap — than the control group (who hadn't taken a nap). What's more, the next day, after a night's sleep, the disparity between the two groups' juggling skills was even greater.[27]

Again, REM sleep is quite useful in consolidating declarative memories associated with emotional content. Yet REM sleep never occurs all by itself: it's always preceded by light and deep SWS, whose combined efforts also serve to consolidate useful information. And, as you will recall, it may well be the frequency of the succession of deep SWS and REM sleep stages that makes for the most effective consolidation. In any case, more and more studies support the hypothesis that we retain more of what we learn, whatever that may be, when learning is followed by a nap providing both deep SWS and REM sleep. A 2003 study by Harvard's psychology department drew attention to this phenomenon: after a session of training in a perceptual task, participants whose nap sleep contained only deep SWS succeeded in stabilising what they'd learned, but those whose naps also produced REM sleep considerably improved their performance on the task concerned.[28]

To sum up, long naps — at least those that, according to the various hypotheses we've just considered, contain light stage-two SWS to consolidate memory, deep stage-three SWS to eliminate parasitic information, and REM sleep to improve memory performance — are ideal for boosting your memory.

So, in the best of all possible worlds, allowing for the distribution of the various stages of nap sleep according to your age, you could adjust the timing and duration of your naps to the kind of learning you wish to consolidate. The nap sleep of infants is rich in deep slow-wave and REM sleep, which is particularly plentiful during the first four months of their lives. Their various daytime naps are crucial to their development, for the plasticity of their brains enables them to learn motor skills, language, the rudiments of social interaction, and a thousand other invaluable skills at a phenomenal speed. A 2015 study by the psychology department at Ruhr University Bochum in Germany with infants aged 6–12 months showed that babies who'd taken a nap of 90 minutes, on average, within four hours of learning novel actions involving declarative memory had learned many more things than those who hadn't napped. And as with the memory of the Japanese student jugglers, the difference was even more pronounced the day after the experiment.[29] Needless to say, this holds true as kids grow older, too: when it comes time to learn language, at about 16 months of age, memorising new words that refer to objects is greatly enhanced by long naps (of about two hours) after learning the new vocabulary. This was demonstrated in a 2015 study by the department of experimental psychology at Oxford University. Sixteen-month-old infants were taught object–word pairs and then divided up into two groups, a 'nap group' and a 'wake group'. Those who took a nap scored significantly higher in a test session two hours later than the ones who didn't have the benefit of a nap.[30]

A 2013 study of children aged three to six by the department of psychological and brain sciences at the University of Massachusetts Amherst reached the same conclusion, though with an important added discovery: when it comes to consolidating declarative memory, the benefit of napping depends on its regularity. In other words, a one-off nap of more than one hour taken by an occasional napper immediately after learning something will be much less effective than the same nap taken by a regular napper.[31]

And don't forget that REM sleep also serves to protect you from painful emotional memories. Faced with any trauma, it may be beneficial to try to reduce the emotional charge by treating yourself now and then to a REM-sleep cure — which only sets in after sleeping for a certain stretch of time (and more quickly in the morning).

BOOSTING MORALE AND INSPIRING DREAMS

Besides the fact that it's bad for your morale to be exhausted, mentally listless, and generally forgetful all the time, a snooze also possesses the priceless power to improve your mood, even in the absence of sleep debt — if only, perhaps, in simply giving you a chance to reclaim 'ownership' of your time, to take a timeout when you feel like it. The reward will be infinitely greater than the time you think you're losing by dozing. Not to mention that a less tired brain will be sharper, more creative and resourceful, more relaxed, and more social.

The 2010 study by Southwest University in Chongqing took advantage of its comparative study of sleeping positions to gauge the effect of napping on the moods of 36 young men and women aged 18–23. Based on a subjective assessment of the subjects' levels of irritability, nervous tension, sadness, and sluggishness, the researchers noted that napping, however comfortable or uncomfortable the conditions, improves these aspects of mood.[32]

A Japanese study conducted three years earlier by the National Institute of Occupational Safety and Health in Kawasaki showed that a 20-minute nap gives more *pleasure* than exposure to bright sunlight for the same amount of time, and is accompanied by a surge of satisfaction and greater serenity.[33] This is, in all likelihood, the reward that appetitive nappers — those who snooze for pleasure — are after.

Sleep makes the most of this 'inner smile'. By freeing up the mind and facilitating hyper-associations, napping serves to rearrange the substrata of memory, a task that our wakeful minds, overwhelmed by incessant thoughts and demands on our attention, never have time to undertake. This enables our minds to access the huge creative potential that lies in our unconscious. To explore that potential, a team at the University of California, San Diego in 2009 assessed the role of nap sleep in creativity using a so-called 'Remote Associates Test' (RAT).[34] Assuming that creativity is the ability to create associations between pieces of information in order to obtain new combinations for a specific purpose (problem-solving), the idea was to test the effect of nap sleep immediately after subjecting participants to such an RAT. It emerged that naps, provided they produce REM sleep, can significantly inspire insight upon waking through the interplay of sleep-specific hyper-associations.[35]

Which is why we owe a number of great discoveries to REM sleep. Among the scientists humble enough to credit the source of their inspirations, the German chemist August Kekulé (1829–1896), the founder of structural organic chemistry, discovered the cyclic structure of the benzene molecule after dreaming of the ouroboros, an ancient ring-shaped symbol depicting a serpent biting its own tail. Likewise, the Austrian biologist Otto Loewi (1873–1961) dreamed (two nights in a row!) of an experiment by which he then succeeded in proving the chemical nature of the transmission of nerve impulses, a hypothesis he'd been trying to verify for 17 years!

Three hacks to stave off inertia

The 20-to-30-minute siesta may be said to be the queen of naps. Certainly, it's the easiest to practise on a daily basis. It has just one drawback: namely, sleep inertia afterwards in the wake of any deep slow-wave sleep. As with longer naps, that consequent sleep inertia can slow down the resumption of activity. But rest assured: there are some tricks you can use to avoid this problem.

SET AN ALARM

First of all, avoiding the risk of deep SWS is a cinch: just take short naps. But how are you to wake up on your own after 10 or 15 minutes of light sleep? Well, if you estimate your sleep-onset latency correctly, limit your sleep time with an alarm clock. Carry out some trials in order to set the countdown correctly, the most important thing being that your naps are short enough not to produce deep SWS.

If, on the other hand, you want to allow yourself a little more nap time but minimise the lethargy caused by deep SWS, be careful not to fall asleep without planning — at least mentally — your wake-up time, for it seems that the body benefits from such anticipation. This is the hypothesis that Jan Born and his team came up with in a 1999 study comparing night sleep when sleepers knew beforehand what time they were going to wake up.[36] The first group thought they'd be allowed to sleep till 9.00 am, but were woken up 'by mistake' at 6.00 am; the second group was also woken up at 6.00 am, but had been forewarned; and the third group was woken up at 9.00 am, the usual wake-up time for all the participants. Remember that morning wake-up is always preceded by the production of stress hormone — dependent on your biological clock — to help you start your day. The experiment showed that, regardless of their biological clocks, stress-hormone production in the sleepers who knew they'd be woken up early commenced as early as 5.00 am. Not so in the sleepers woken up 'by mistake'. In short, this hormone is capable of anticipating wake-up if the hour is known ahead of time, which affords us some interesting prospects for napping purposes. So don't hesitate to fix your wake-up time in advance and thereby set your inner alarm clock.

Experienced practitioners, who've accustomed their bodies to their habitual nap duration, generally don't need noisy alarms. Then again, if ever they wish to vary that duration — from, say, a half-hour to a ten-minute catnap, for example — a ringing alarm clock may prove useful.

THE COFFEE NAP

One trick that doesn't occur to very many people is to drink some coffee just before closing your eyes: caffeine, which takes about 20 minutes to kick in, will dispel sleep inertia after a nap of 20 minutes or more.[37] Caffeine possesses this ability to counter the effects of drowsiness because it's chemically similar to adenosine, which is the residue of the breakdown of ATP (adenosine triphosphate), the energy-carrying fuel used by the cells of all living things to synthesise the substances that nature requires them to produce. However, it's this very adenosine that causes us to switch to sleep mode when it accumulates to a maximum threshold. Caffeine binds to the same brain receptors as adenosine, thus preventing it from binding to them and performing its natural function of slowing down neural activity. It takes four to six hours for half of the caffeine consumed to be eliminated by the brain, and its stimulant effect is cumulative: so if you have some coffee at 9.00 in the morning, the second half of the caffeine will kick in between

1.00 and 3.00 pm — at the very moment you might be downing a second dose to finish off your lunch. Bear this in mind as you drink one cup after another during the day: ideally, caffeine should be gone and out of your system by the time you turn in.

As we've seen, combining caffeine and a nap has the effect of improving cognitive performance and subjective alertness more than either practice does by itself. It may also have the advantage, as discussed below, of enhancing the analgesic properties of napping. Again, though, make sure to allow for the enduring effects of caffeine: you shouldn't combine coffee with a late nap, around 3.00 or 4.00 pm, unless you're gearing yourself up to work that night.

FROM BED TO BATHING ... IN LIGHT

If you don't like coffee (or tea, which contains caffeine, too), or if you're caffeine intolerant or wary of abusing it, there is an alternative: light-therapy lamps (ranging from 100 to several thousand lux). Just as caffeine blocks the action of adenosine, light — of any wavelength — inhibits the secretion of melatonin, which, as you now know, obeys the body clock and primes you every evening for your night-time z's. Light isn't helpful when you're getting ready to sleep — especially, and this is worth reiterating, the short waves of blue light emanating from the screens of our computers, tablets, and mobile phones. On the other hand, its action on certain photoreceptors in the retina can improve vigilance, cognitive performance, and mood.[38]

With my team at Hôtel-Dieu Hospital, I recently measured the effects of acute sleep restriction (two consecutive nights of only three hours' sleep) on young adults (19–33 years old) by following them in real-life conditions, in their day-to-day lives (which are more stressful than conditions in the lab). We found that repeated exposure (four times a day, in the morning and early, mid, and late afternoon) to low-intensity blue light (100 lux) enabled them to recover a high level of alertness and cognitive performance and restore the balance of their stress systems even without a nap.[39] Naturally, combining this light therapy with a nap would have greatly increased its beneficial effects.

In 2003, the department of behavioural sciences at Hiroshima University demonstrated the equivalence, in terms of subjective alertness, between the combination of a nap and a (pre-nap) double espresso and that of a nap and (post-nap) exposure to intense (2,000 lux) white light.[40] The advantage of blue light over white light is plain to see: it can be of much lower intensity and yet prove equally or even more effective. But if it's used too intensively, it can pose a greater risk to the cellular integrity of the retina.[41]

At any rate, napping is the only effective means of countering sleep debt, the only one that enables you to recover and to dispel sleep pressure. Caffeine and light, whatever benefits they may have to offer, merely stimulate our arousal systems and performance. In other words, they treat the symptoms, not the ailment. Used in moderation, however, they are excellent aids in dissipating sleep inertia after a recovery nap.

7

A Therapeutic Stroll

The difference between a good night's sleep and a siesta is like the difference between a hike and a stroll: the former takes time and requires preparation, good starting conditions, and clear skies; the latter can be improvised, done under cloudy skies, and even slipped in between storms. You'll get more air and exercise from a hike, and your body will remember it for a longer time, yet you'll still benefit from a breath of fresh air and stretching your legs for a while during a stroll. Similarly, a nap provides all the benefits to be had from the physiological functions of a night's sleep, only on a smaller scale. When you nod off, your inner medical team goes into action, albeit providing somewhat reduced service, and the show starts again: a ballet of hormones, a release of tension, the music of memories, and, if the nap is long enough, some impromptu dreams. In light of the latest scientific discoveries, let's see how this little day clinic pulls off the feat of providing the daily care that our nights don't have enough time to administer. Most importantly, let's see what napping can do about the health risks of sleep debt.

An antidote to drowsiness

Staying awake for too long will sooner or later undermine our cognitive abilities and has the potential to put us in danger: on the road, at home, or at work. With diminished alertness and a lack of responsiveness to an unexpected stimulus, you're at risk of an accident. Laboratory studies have shown that naps are effective in boosting sustained alertness and that the production of deeper sleep can have lasting cognitive benefits. So we ought to benefit from napping in our day-to-day lives, especially before undertaking a long drive or working at night in a state of sleep debt.

A team at the Bordeaux Sleep Clinic verified these benefits in real-life conditions on a motorway in south-west France.[1] They tested two groups of 12 drivers each — those in one aged 20–25, those in the other aged 40–50 — driving 200 kilometres at a speed of 130 km/h under favourable weather conditions between 2.00 and 3.30 in the morning. To gauge their alertness, the researchers simply counted the number of unintended lane departures: how many times they crossed the white lines. It turned out that the drivers who'd taken a nap of about 20 minutes at 1.00 am or taken 200 mg of caffeine (the equivalent of two espressos) half an hour before departure had a significantly smaller number of driving errors.

It's worth noting that the nap had a far more beneficial effect on the younger drivers than on the older ones, who, on the other hand, benefited more from the coffee. This was probably because the older drivers were unable to sleep as long as the younger ones in the 30 minutes allotted (they got an average of 14 minutes' sleep, versus 22 minutes for the younger drivers) and therefore less of a boost from the delta waves of deep slow-wave sleep. They'd have been better off with a slightly longer nap, but, as you now know, sleep pressure becomes less efficacious with age. In any case, the figures speak for themselves.

In 2004, the University of Genoa's Centre of Sleep Medicine carried out a study of police officers who'd been working shifts for more than five years. Some of them had caused traffic accidents during their patrols: the study revealed that the absence of a nap before going on duty had increased the likelihood of such accidents by 38 per cent.[2]

Painkiller

Everyone knows what it's like when you don't get a good night's sleep and then feel more sensitive to pain the next day, whereupon the ensuing aches and pains keep you from getting a good night's sleep the next night, too. If you have a headache or backache, if your joints or other parts of your body hurt and you can't get enough sleep at night, I can't impress upon you enough the importance of taking a nap.

The study that I conducted with Serge Perrot's team revealed that, after acute sleep deprivation (two hours of sleep between 2.00 am and 4.00 am), a half-hour nap (50 per cent deep slow-wave sleep) was sufficient to restore normal pain sensitivity, i.e. the way we feel after sleeping properly at night.[3] The hypersensitivity due to lack of sleep varied according to the cause of the pain and the anatomical area concerned: pain in the lower back manifested itself more rapidly when caused by increasing heat than by mechanical pressure; the opposite obtained for pain in the shoulder muscles. But after the nap, there was no observable difference between these two areas, and pain sensitivity had reverted to its equilibrium level.

Napping might also turn out to be useful in relieving pain caused by fibromyalgia. A syndrome that chiefly affects women, fibromyalgia, according to the medical dictionary of the French National Academy of Medicine, 'combines diffuse pain in muscles, joints, and especially tendons with acute fatigue, especially in the morning, sleep and mood impairment, and minor cognitive disorders'. Alice Theadom's 2015 study of 1,044 fibromyalgia sufferers (92.5 per cent women) found that napping too frequently (daily) and, above all, for too long (an hour or more) ended up worsening the symptoms — probably simply because chronic napping is a sign of overall physical deterioration — but that short naps (half an hour or less), on the contrary, could have an analgesic effect.[4]

Yet we shouldn't write off long naps as a means of alleviating other types of pain. The Japanese National Institute of Occupational Safety and Health examined the relation between musculoskeletal pain in 54 nurses (average age: 32) and the frequency of naps during their eight-hour night shifts. The study found that pain in their arms and legs was reduced by factors of three and two, respectively, in nurses who took at least one nap of 76 minutes (on average) every other work night compared to nurses who took fewer naps or none at all.[5]

Generally speaking, it's hard to treat chronic pain pharmacologically in the long term, mainly owing to the decline in drug effectiveness over time and the risk of addiction. So you may well be better off taking regular half-hour naps to cut down on your long-term consumption of painkillers. One reason is that how we perceive and withstand pain depends in part on our level of alertness. In 2017, a Harvard Medical School team found that conventional analgesics (such as ibuprofen) were far less effective in reducing pain responses in sleep-deprived mice than psychoactive, 'wake-promoting' (mainly caffeine-based) drugs administered to diminish drowsiness.[6]

A calming balm

Lack of sleep causes stress, which, in turn, causes insomnia and, consequently, even more stress. This cycle eventually wears the body out and plunges the insomniac into depression.[7] Hence the vital importance of finding an alternative before you end up adding antidepressants to painkillers and sleeping pills, all of which should only be a last resort. If you're a stressed-out insomniac, give some thought to your lifestyle (physical activity and diet) and don't scoff at the benefits of napping. Above all, don't imagine it will make it any harder to get to sleep at night. On the contrary, a short nap (20 minutes) will improve your mood and — if taken at least six hours before bedtime — your propensity to fall asleep and stay asleep at night.[8]

We're all familiar with stress, which is constant; what varies is its intensity and our ability to channel it. Over the course of a normal day, the level of stress hormones (cortisol and catecholamines such as adrenaline) fluctuates regularly according to the rhythm of our biological clock: it peaks early in the morning (around 8.00–10.00 am), decreases until early and mid-afternoon, then stabilises and/or increases later in the afternoon (around 5.00 pm), gradually coming back down to hit rock bottom in the middle of the night.[9] But why does it drop off after lunchtime, just when we need plenty of vim and vigour to rev us up again in the afternoon? Maybe for the very reason that our bodies are designed in such a way that we need to take breaks, even at midday, particularly to slow down the activity of our principal hormonal response to stress. Especially seeing as positive stress is often admixed, sometimes daily, with negative stress: fatigue, anxiety, anger, pressure, overwork ...

As I said, a small dose of stress is essential for getting up on the right foot in the morning. A larger dose, though of limited duration, can be extremely stimulating in order to complete an undertaking. Chronic, non-stop, suffocating stress, however, has nothing but disastrous effects on the body. In the long run, once again, it can even increase the likelihood of a heart attack or stroke. And we've yet to find a more natural remedy for stress than the regular practice of napping.

The first American experiment to show the effects of napping on stress levels was carried out in 2007 by the department of psychiatry at the University of Pennsylvania. It revealed that, for healthy adults aged 18–30, a two-hour nap between 2.00 pm and 4.00 pm after a night of total sleep deprivation reduced the level of stress hormones in the blood by about one-third.[10] In a word, the two-hour nap served as a partial substitute for a full night's sleep. But is there a less time-consuming way to regulate stress?

I teamed up with colleagues at the Free University of Brussels in 2011 and then again at Hôtel-Dieu Hospital in 2015 to find out. Using two different protocols, we analysed the effects of a 30-minute nap on healthy young adults after a night of sleep restriction (two hours' sleep from 2.00 am to 4.00 am).[11] The subjects were 18–25 years old in the first study and 25–32 years old in the second. In both cases, to avoid the invasive method of taking blood samples, we instead regularly measured cortisol levels in their saliva (in both studies) and catecholamine levels in their urine (in the second study). In the first study, some of the subjects took a half-hour nap at 1.00 pm, whereas the second study involved two half-hour naps, one at 9.30 am and another at 3.30 pm. In each case, the concentration of stress hormones in the body plummeted by about half after the naps.

Further studies will eventually tell us about the effects of shorter naps (five, ten, or 15 minutes) on stress. What we know today is that after a night of four hours' sleep, the first five minutes of a nap significantly lower blood pressure and heart rate (both of which, as you may recall, require stress hormones to increase).[12] On the other hand, a close correlation has been discovered between deep slow-wave sleep (SWS) during daytime sleep and the level of the stress hormone cortisol: they're inversely proportional. The greater the oscillation of deep SWS waves, the lower the cortisol levels, and vice versa.[13] While stress-hormone levels decline slowly during the first part of the night (rich in delta waves), they drop precipitously during a nap as soon as deep SWS sets in.

It stands to reason that a five-minute nap, which doesn't last long enough to get beyond the stage of light SWS, may be insufficient to significantly alleviate stress. The longer your nap (between half an hour and one hour), the better your chances of getting some deep SWS and abating your stress-hormone levels. And in these stressful times, it would be a shame to pass up that opportunity. In 2007, a team of researchers in the department of hygiene, epidemiology, and medical statistics at the University of Athens School of Medicine, after following up a population of a little over 23,000 adults aged 20–86 for over six years, found that the prevalence of cardiovascular events was 37 per cent lower among all those who'd taken early-afternoon siestas of roughly 30 minutes at least three times a week during the period covered.[14]

Keeper of the immunological peace

When you're injured or contract a disease, your internal systems shift into an acute stress configuration. Effective protection requires rapid deployment of the appropriate white blood cells — lymphocytes, phagocytes, granulocytes — to the injured or infected areas of the body. Just as stress hormones are permanently present in the blood in some amount, our immune system also provides non-stop service, which varies according to our biological clocks and our sleep: during the day, when cortisol and adrenaline levels peak, some lymphocytes and phagocytes reach high levels, too; at night, when cortisol levels are low, lymphocyte production is activated based on our body's immunological memory.[15]

Therefore, if sleep is restricted, the levels of certain immune cells will be higher than they should be, as demonstrated by two studies carried out at the Free University of Brussels in 2007 and 2009. Each study involved various protocols: the first two protocols restricted sleep for groups of women aged 55–65 and men aged 21–28 to only four hours for three nights; the third restricted a group of 22-to-29-year-old men to five hours' sleep for five nights in a row, as in a busy working week; and the fourth restricted another group of 18-to-27-year-old men to two hours' sleep (2.00 am–4.00 am) for one night, followed by a half-hour early-afternoon nap (1.00 pm–1.30 pm).[16] It immediately became clear that the greater the sleep restriction, the higher the neutrophil count. Neutrophils — which were present in overabundance despite the absence of infection — are phagocytes that tend to release oxidising and proteolytic agents that can damage vascular walls, as evidenced in a 2006 Finnish study of 219 men in their 50s.[17] A follow-up study spanning 44 years (1958–2002) by the National Institutes of Health in Baltimore, Maryland, found that the risk of death from cardiovascular disease is about 33 per cent higher among both men and women whose neutrophil counts in the blood (even if only slightly higher than normal) steadily increase with age.[18]

Insufficient recovery sleep is oftentimes the cause. As with oxidative stress, one night of recovery won't suffice for a return to status quo in cases of acute sleep debt. Some studies have delved deeper here. After a night of eight hours' recovery sleep (following a night of only two hours' sleep[19] or, in another experiment, continuous sleep deprivation for 64 hours),[20] or after seven nights of recovery sleep (following five consecutive nights of five hours' sleep),[21] the participants' neutrophil counts still hadn't returned to their normal values.

My 2011 study at the Free University of Brussels found that after a night of only two hours' sleep, it takes a ten-hour night's sleep to revert to a standard neutrophil count. But the most interesting discovery was that a half-hour nap yielded the same result.[22] The nap-induced drop in catecholamines and cortisol precipitated a concomitant drop in neutrophils. And what goes for immune cells goes for pro-inflammatory molecules, too: a long two-hour nap in the early afternoon lowers abnormally elevated levels of interleukin-6 in the blood.[23] Conversely, sleep deprivation reduces interleukin-6 levels in saliva, although it's essential for neutralising infectious agents because, besides its pro-inflammatory functions, interleukin-6 also promotes the production of antibodies by lymphocytes.[24] It turns out that a 30-minute nap, half of which comprises deep SWS, is capable of bringing interleukin-6 levels in saliva back up to standard values.[25]

Another case in point is a study by a team in the neuroscience and behaviour graduate program at the University of Massachusetts Amherst. The study population comprised a cohort of 2,147 young adults (average age: 29), including short, medium, and long sleepers, who had taken naps 'every day' (3.5 per cent), 'almost every day' (7.3 per cent), or 'never to a few times' (89.2 per cent) over the previous seven days. Their blood levels of C-reactive protein (CRP), which is considered a reliable biological marker of inflammation, spoke volumes: among the short sleepers (five hours or less a night), those who'd napped almost every day had lower average values of CRP than the non- or very occasional nappers — but also significantly lower than the chronic nappers, whose long daily naps, once again, indicated a general decline in their state of health.[26]

More and more experimental data confirms that chronic sleep restriction can lead to a low-grade inflammatory process,[27] so there's every reason to believe that the *short-term* benefits of napping to combat the stress and immunological and inflammatory processes brought on by sleep debt also have the *long-term* effect of protecting against cardiovascular risk.

Heart nurse

If there's one statistic worth retaining from the myriad studies in the world of sleep research and the multitude cited in this book, it's this: more than a third of cardiovascular events could be prevented if the denizens of this planet took just a little time out for a nap.

Back in the 1980s, some Greek scientists at the University of Athens School of Medicine underscored the virtues of a siesta: although based on a sample of only 200 people, their study concluded that a nap reduces our cardiovascular risk by about one-third.[28] More recently, the same medical school confirmed that regular napping — for at least half an hour three times a week — reduces the risk of coronary mortality by one-third, too. This time around, the study followed up more than 23,000 Greeks aged 20–86 over the course of six years. The men and women in the cohort were selected to control for comorbidity factors like smoking, excess weight, sedentary lifestyle, level of education, type of diet, and so on that were liable to confound the results.[29]

Regular or occasional napping was found to have an even more protective effect on men who were gainfully employed — and therefore potentially sleep-deprived during the week: it halved their risk.

Over in Spain, at the University of Navarra, a six-year longitudinal study of about 10,000 men and women aged 20–90, with no chronic diseases or obesity at baseline (that is, at the time of selection), found that the risk of becoming obese is twice as high for short sleepers (five hours of sleep or less a night) as for medium sleepers (seven to eight hours of sleep a night). Above all, the risk is one-third lower for those who take a half-hour daily nap, whether they're short, medium, or long sleepers.[30] And the icing on the cake is that those who took longer naps didn't derive any greater benefits.

What holds for the Mediterranean seems to hold elsewhere in the world as well. In the neurosurgery department of a university hospital in Beijing, a recent study of several hundred patients with intracerebral aneurysms found that two-thirds of them were regular (short or long) nappers, and this wholesome habit actually halved their risk of suffering a ruptured aneurysm.[31] In fact, comparing the results of all the studies around the world on the connection between napping and cardiovascular risk — including Tokyo University's 2015 meta-analysis of more than 150,000 participants, mostly over 50 years of age (60 per cent women), over an average follow-up period of 11 years — *taking a daily nap for 10–30 minutes significantly reduces the risk of cardiovascular disease*, regardless of the sleeper's age.[32]

A mere 10–30 minutes a day ... It's worth it, isn't it?

Conclusion

If I've given so much importance to long night-time sleep before delving into short day-time sleep, it's because you can't understand the virtues of the siesta without keeping in mind the benefits of a good night's sleep. Napping has the attributes, though not the stature, of night-time sleep. It's a night's sleep in miniature. It's only a backup, never a substitute. First, we've got to make sure we're getting our fill of nightly slumbers. Only then can we offer our physical and cognitive capacities, vitiated by long-term involuntary sleep deprivation, the privilege of being partially, if not fully, restored by napping. If you feel a lack of sleep is undermining your health, always start by taking control of your nights again: try to get to bed an hour earlier and respect your sleep windows and your real sleep needs.

To be sure, sleeping doesn't really have a better reputation than napping, especially in the working world. We need to change our view of sleep and put paid to the misconception that sleeping's a waste of time, opportunities, and money. 'I'll have plenty of time to sleep when I'm dead and gone' is idle and inane bluster. Sooner or later, the dauntless daredevils of sleep deprivation will go too far and find themselves hoist with their own petard, ensnared by all the health hazards we've just enumerated. Your sleep is your health, and napping its salutary supplement.

So don't trust new technologies that would lead you to believe you can optimise sleep so as to be more and more active and productive during your — extended — waking hours. These schemes are often predicated on the wrongheaded assumption that there's an interest in sleeping less rather than trying in vain to get as much sleep as you need. We now have newfangled watches and other connected devices for the sole purpose of tracking our sleep: primarily sleep duration, the number of awakenings, and percentage breakdowns of the various sleep stages. These 'tools' are a fun way to familiarise ourselves with the 'little train' of sleep, as I've called it, but they often deliver dubious data, based solely on our heart rate and physical movements — or lack thereof. Not only that, but we now have processes designed to interact with the brain to optimise sleep by such enhancements as soothing music, the sound of rainfall in the Amazon, red light, random words purported to help put you to sleep, sound stimuli to improve sleep quality, and blue light accompanied by gentle birdsong to help you rise and shine in the morning. All too frequently, the only certainty these innovations have to offer is their price, which can be exorbitant. 'Smart' mattress toppers, futuristic designer 'sleeping pods' for a 'regenerative nap', 'siesta bars', reclining 'zero-gravity' chairs, virtual-reality 'relaxation helmets', you name it — all these businesses besieging the last remaining stronghold of our private lives that has yet to be mined for profit: our sleep, which never asked for all that paraphernalia.

Underlying all these offers is the myth that wakefulness is the be-all and end-all, a credo that largely inspires the modern economy. It's a delusion that consists in believing we can circumvent the rhythms of our natural sleep–wake cycles by means of technoscience and reduce our needs to the bare minimum by optimising them, as if we were mere machines. This is bound to benefit a handful of merchandise vendors, but I doubt it's going to improve our quality of life, our health, and our creativity. At most, it may slake our servile appetite for the inexhaustible innovations of the connected world. So let's preserve our last bastion of freedom. Sleep doesn't need reinventing. It needs to be understood, respected, and protected.

Polyphasic sleep, whose merits are widely vaunted, and which involves fragmenting our sleep over the course of each day in order to be more and more productive and successful, partakes of this cult of the modern hero, the 'achiever'. At most, we can endorse biphasic sleep, which boils down to nothing more than combining night-time sleep, though regrettably cut short (by one or two hours, depending on how hard that is for each sleeper), with a recovery siesta of 20–30 minutes in the early afternoon. This is precisely the idea I have championed here. But not a single study to date shows that this sleep schedule is better than the luxury of a full night's sleep.

What is certain, on the other hand, is that taking a nap every day to offset our sleep debt is a natural and beneficial medication. From morning to mid-afternoon, the variety of possible nap durations, from 10–90 minutes, depending on our availability, age, and needs, makes it a judicious and formidable weapon for reinforcing not only our metabolic, hormonal, immune, somesthetic, and cardiovascular functions, but also our alertness, cognitive performance, memory, mood, empathy, and creativity.

Like sleeping, napping tells us about time and our relationship to time in a world which is, as a rule, hell-bent on stealing it from us. Taking a nap is taking possession of yourself a little in order to construct and rebuild yourself, day after day, to live in better health and with a sharper, more inventive, and consequently freer mind. That is why we must restore our sleep balance without further ado. Let us heed the words of the late, great French philosopher Alain, whose exhortation, like a Socratic motto, also serves as the epigraph to this book: *Sachons dormir, nous saurons veiller*. Freely translated: 'Let's learn how to sleep, then we'll know how to be wakeful.' Sleep is more than a matter of public health, it's a social issue of vital importance to an enlightened democracy. For we need to sleep if we're to have any hope of thinking clearly.

The Call
of the Siesta

The Napper's Cheat Sheet

To each their own nap!

This little 'cheat sheet' is based on scientifically obtained statistical averages, for the most part involving cases of sleep deprivation.

The effects of a nap depend on too many variables to constitute an exact science. Your age, state of health, sleep debt, and degree of nervousness; the timing, duration, and frequency of your naps; the quality of your armchair or bed, etc. – are all factors with a bearing on the results of these micro-cures for sleep debt.

That said, these suggestions should prove helpful in most cases. So make the most of them!

(NB: The nap durations indicated in the following correspond to the time spent actually asleep.)

Golden rules

- Listen to your body and heed its calls for sleep without delay: waiting too long will make it harder to fall asleep.
- Concentrate on your breathing in order to relax and get to sleep more quickly. Basically, chill out.
- Banish all sources of blue light: switch off the screens of your electronic devices (computers, tablets, smartphones).
- Find a dark, quiet spot for your nap.
- The more naps you take, the easier it gets. It will become a pleasant and pleasurable habit.

What's the best napping position?

- The ideal position is lying flat on a bed or sofa.
- If a bed or sofa isn't available, sleep seated in an armchair. If possible, tilt the backrest at least 40 degrees (relative to upright position) and use a neck cushion to support your neck.

10-minute nap

BENEFITS

- You're guaranteed to feel refreshed afterwards.
- Effective five minutes after wake-up: ideal before an important meeting or appointment.
- Highly effective in combating drowsiness and stimulating cognitive performance for a short while (two hours).

- Ideal for rapid recovery from a temporary slump in physical and/or intellectual capacities.

LIMITATIONS

- Not very effective for a longer duration: in cases of considerable sleep debt and chronic fatigue, this short nap won't suffice to brace you for an extended period of wakefulness.

20- to 30-minute nap

BENEFITS

- Alleviates pain, stress; boosts immunity; reduces risk of illness and cardiovascular events.
- Stimulates alertness for over two hours.
- Improves mood.

LIMITATIONS

- May cause sleep inertia (grogginess upon waking): the longer the deep slow-wave sleep (SWS) stage, the more inertia you're likely to experience.
- Effects don't kick in till 20–30 minutes later.
- Takes at least 30–40 minutes' time (allowing for sleep-onset latency – the time it takes to fall asleep).

One- to two-hour nap

BENEFITS

- Ideal for repaying sleep debt faster.
- Effective preparation for night work: significantly improves alertness and cognitive performance for at least five hours after waking.
- Helps consolidate memory and boost creativity and immunity.

(NB: A 90-minute nap corresponds to a complete sleep cycle, including all four stages.)

LIMITATIONS

- Sleep inertia after wake-up.
- In some older people, long naps may conflict with sleep–wake rhythms and may be associated with cognitive decline and increased cardiovascular risk.

To boost your creativity

- Take naps of more than one hour, which provide some REM sleep. Or take shorter naps in the morning.

To reduce your stress and boost your immune system

- Take a siesta of 30 minutes to one hour, preferably in the early afternoon. The more deep slow-wave sleep (SWS) you get, the more pronounced the effects will be.

To reduce cardiovascular risk

- Take 30-minute naps, preferably midday/early afternoon (between noon and 4.00 pm).

The best times for a nap

- Morning: between 9.00 am and noon, to recoup some of the REM sleep lost by interrupting your night sleep too early.
- Midday/early afternoon: between noon and 4.00 pm for a 10- to 30-minute siesta.
- In the early afternoon, a siesta of over 30 minutes will be richer in deep slow-wave sleep (SWS) than REM sleep for early birds (morning chronotype), but richer in REM sleep than deep SWS for night owls (evening chronotype).
- After 5.00 pm, don't take a nap of more than 20 or 30 minutes if you're planning to turn in for the night around 11.00 pm: leave an interval of at least six hours between your afternoon nap and bedtime.

To wake up refreshed

- Set your alarm clock to wake you up after 20 minutes to avert the risk of plunging into deep short-wave sleep (SWS).
- Drink a cup of coffee just before a 20-minute nap: it's guaranteed to do the trick.
- Use a light-therapy lamp to activate your body's arousal systems and reduce sleep inertia.

Recap: the stages of sleep

Nap sleep, like night sleep, is divided up into four stages, which determine its restorative properties:

- Light slow-wave sleep (SWS) (stages one and two): plentiful at the beginning of a nap and characteristic of short naps of five to ten minutes; provides an immediate boost of energy.
- Deep SWS (stage three): sets in after about 20 minutes of sleep (or earlier in case of substantial sleep debt); the most restorative kind of sleep, but also produces the heaviest sleep inertia after wake-up (making for a longer transition between sleep and wakefulness).
- REM sleep: present in naps of more than one hour (or less in the morning); boosts creativity and memory and produces vivid dreams.

Acknowledgements

I'd like to warmly thank my editor, Sylvie Fenczak, for her attentiveness and enthusiasm throughout the development of this book.

I'd also like to give a cordial 'shout-out' to Daniel Hantaï, my thesis supervisor and the first mentor in my research career.

My gratitude goes to Myriam Kerkhofs and Karim Zouaoui Boudjeltia at the Sleep Laboratory of André Vésale Hospital and the Free University of Brussels, who, in a context of concerted creative energy, gave me the confidence necessary to narrow down my research and follow my intuitions as far as they might take me.

My scientific thought was also enriched by the many exchanges I had with other researchers within the Marie Skłodowska-Curie Actions research network during my stints at various sleep and chronobiology research laboratories (in Guildford, Helsinki, Zurich, Lyon ...). With their different cultures but shared interests, all these researchers infused me with their tremendous open-mindedness, for which I am grateful.

I'd like to thank the members of the Hôtel-Dieu Sleep Centre and our VIFASOM research team at Paris Descartes University for their support and commitment, especially my colleagues Damien Léger, Mounir Chennaoui, Maxime

Elbaz, Nicolas Hommey, Marie-Francoise Vecchierini, and Thomas Andrillon. Day after day, we seek to develop a true spirit of innovation and freedom.

Last but not least, thanks so much to all those who have contributed to this field of endeavour, without whom the research presented in this book would have remained ... *dormant!*

Bibliography

Adrien J., 'Neurobiological bases for the relation between sleep and depression', *Sleep Med. Rev.*, 6(5), Oct. 2002, pp. 341-351.

Adrien J., 'Dormir? Comment et pourquoi', in S. Royant-Parola (ed.), *Les Mécanismes du sommeil. Rythmes et pathologies*, Le Pommier, 2007.

Aeschbach D., Postolache T. T., Sher L., Matthews J. R., Jackson M. A., Wehr T. A., 'Evidence from the waking electroencephalogram that short sleepers live under higher homeostatic sleep pressure than long sleepers', *Neuroscience*, 102(3), 2001, pp. 493-502.

Aeschbach D., Sher L., Postolache T. T., Matthews J. R., Jackson M. A., Wehr T. A., 'A longer biological night in long sleepers than in short sleepers', *J. Clin. Endocrinol. Metab.*, 88(1), 2003, pp. 26-30.

Åkerstedt T., Bassetti C., Cirignotta F., García-Borreguero D., Gonçalves M., Horne J., Léger D., Partinen M., Penzel T., Philip P., Verster J. C., *La Somnolence au volant. Livre blanc*, ASFA, INSV, 2013.

Alain, *Propos I*, 'Persuasion' [1921], Coll. 'Bibliothèque de la Pléiade', Gallimard, 1956, pp. 330-332.

Alexandre C., Latremoliere A., Ferreira A., Miracca G., Yamamoto M., Scammell T. E., Woolf C. J., 'Decreased alertness due to sleep loss increases pain sensitivity in mice', *Nat. Med.*, 23(6), June 2017, pp. 768-774.

Allan J. S., Czeisler C. A., 'Persistence of the circadian thyrotropin rhythm under constant conditions and after light-induced shifts of circadian phase', *J. Clin. Endocrinol. Metab.*, 79(2), Aug. 1994, pp. 508-512.

Allebrandt K. V., Amin N., Müller-Myhsok B., Esko T., Teder-Laving M., Azevedo R. V., Hayward C., Van Mill J., Vogelzangs N., Green E. W., Melville S. A., Lichtner P., Wichmann H. E., Oostra B. A., Janssens A. C., Campbell H., Wilson J. F., Hicks A. A., Pramstaller P. P., Dogas Z., Rudan I., Merrow M., Penninx B., Kyriacou C. P., Metspalu A., Van Duijn C. M., Meitinger T., Roenneberg T., 'A K(ATP) channel gene effect on sleep duration: from genome-wide association studies to function in Drosophila', *Molecular Psychiatry*, 18(1), Jan. 2013, pp. 122-132.

ANSES 2018: https://www.anses.fr/en/system/files/NUT2016SA0209.pdf.

Arnulf I., Grosliere L., Le Corvec T., Golmard J. L., Lascols O., Duguet A., 'Will students pass a competitive exam that they failed in their dreams?', *Conscious. Cogn.*, Oct. 2014, pp. 36–47.

Aubert C., 'Longévité, les limites d'une espérance', *Le Monde diplomatique*, June 2018.

Axelsson J., Ingre M., Åkerstedt T., HolmbÄck U., 'Effects of acutely displaced sleep on testosterone', *J. Clin. Endocrinol. Metab.*, 90(8), Aug. 2005, pp. 4,530–4,535.

Ayas N. T., White D. P., Manson J. E., Stampfer M. J., Speizer F. E., Malhotra A., Hu F. B., 'A prospective study of sleep duration and coronary heart disease in women', *Arch. Intern. Med.*, 163(2), Jan. 2003, pp. 205–209.

Banks S., Dinges D. F., 'Behavioral and physiological consequences of sleep restriction', *J. Clin. Sleep Med.*, 3(5), Aug. 2007, pp. 519–528.

Basner M., Spaeth A. M., Dinges D. F., 'Sociodemographic characteristics and waking activities and their role in the timing and duration of sleep', *Sleep*, 37(12), Dec. 2014, pp. 1,889–1,906.

Baum K. T., Desai A., Field J., Miller L. E., Rausch J., Beebe D. W., 'Sleep restriction worsens mood and emotion regulation in adolescents', *J. Child Psychol. Psychiatry*, 55(2), 2014, pp. 180–190.

Bayon V., Leger D., Gomez-Merino D., Vecchierini M. F., Chennaoui M., 'Sleep debt and obesity', *Ann. Med.*, 46(5), Aug. 2014, pp. 264–272.

Beck F., Gautier A., Guignard R., Richard J.-B. (eds.), *Baromètre santé 2010. Attitudes et comportements de santé*, INPES, Saint-Denis, 2012.

Benedict C., Scheller J., Rose-John S., Born J., Marshall L., 'Enhancing influence of intranasal interleukin-6 on slow-wave activity and memory consolidation during sleep', *FASEB J.*, 23(10), Oct. 2009, pp. 3,629–3,636.

Besedovsky L., Lange T., Born J., 'Sleep and immune function', *Pflugers Arch.*, 463(1), Jan. 2012, pp. 121–137.

Besedovsky L., Ngo H.-V. V., Dimitrov S., Gassenmaier C., Lehmann R., Born J., 'Auditory closed-loop stimulation of EEG slow oscillations strengthens sleep and signs of its immune-supportive function', *Nature Communications*, 8, art. 1984, Dec. 2017.

Billiard M., *Le Sommeil et l'Éveil*, Masson, 2000.

Billiard M., Dauvilliers Y. [2005], *Les Troubles du sommeil*, Masson, repr. 2011.

Blanco R. A., Ziegler T. R., Carlson B. A., Cheng P. Y., Park Y., Cotsonis G. A., Accardi C. J., Jones D. P., 'Diurnal variation in glutathione and cysteine redox states in human plasma', *Am. J. Clin. Nutr.*, 86(4), Oct. 2007, pp. 1,016–1,023.

Born J., Hansen K., Marshall L., Mölle M., Fehm H. L., 'Timing the end of nocturnal sleep', *Nature*, 397(6714), Jan. 1999, pp. 29–30.

Boudjeltia K. Z., Faraut B., Stenuit P., Esposito M. J., Dyzma M., Brohée D., Ducobu J., Vanhaeverbeek M., Kerkhofs M., 'Sleep restriction increases white blood cells, mainly neutrophil count, in young healthy men: a pilot study', *Vasc. Health Risk Manag.*, 4(6), Dec. 2008, pp. 1,467–1,470.

Boudjeltia K. Z., Faraut B., Esposito M. J., Stenuit P., Dyzma M., Van Antwerpen P., Brohée D., Vanhamme L., Moguilevsky N., Vanhaeverbeek M., Kerkhofs M., 'Temporal dissociation between myeloperoxidase (MPO)-modified LDL and MPO elevations during chronic sleep restriction and recovery in healthy young men', *PLoS One*, 6(11), Nov. 2011.

Bouscoulet L. T., Vázquez-García J. C., Muiño A., Márquez M., López M. V., Oca M. M. de, Talamo C., Valdivia G., Pertuze J., Menezes A. M., Pérez-Padilla R., 'Prevalence of sleep related symptoms in four Latin American cities', *J. Clin. Sleep Med.*, 4(6), Dec. 2008, pp. 579–585.

Bremner W. J., 'Testosterone deficiency and replacement in older men', *N. Engl. J. Med.*, July 2010, 363(2), pp. 189–191.

Brescianini S., Volzone A., Fragnani C., Patriarca V., Grimaldi V., Lanni R., Serino L., Mastroiacovo P., Stazi M. A., 'Genetic and environmental factors shape infant sleep patterns: a study of 18-month-old twins', *Pediatrics*, 127(5), May 2011, pp. 1,296–1,302.

Brooks A., Lack L., 'A brief afternoon nap following nocturnal sleep restriction: which nap duration is most recuperative?', *Sleep*, 29(6), June 2006, pp. 831–840.

Brygo J., 'Le routier américain, une icône en voie de disparition', *Le Monde diplomatique*, Aug. 2018, p. 5.

Cai D. J., Mednick S. A., Harrison E. M., Kanady J. C., Mednick S. C., 'REM, not incubation, improves creativity by priming associative networks', *PNAS*, 106(25), June 2009, pp. 10,130–10,134.

Cajochen C., 'Alerting effects of light', *Sleep Med. Rev.*, 11(6), Dec. 2007, pp. 453–464.

Cao Z., Shen L., Wu J., Yang H., Fang W., Chen W., Yuan J., Wang Y., Liang Y., Wu T., 'The effects of midday nap duration on the risk of hypertension in a middle-aged and older Chinese population: a preliminary evidence from the Tongji-Dongfeng Cohort Study, China', *J. Hypertens.*, 32(10), Oct. 2014, pp. 1,993–1,998.

Cappuccio F. P., Miller M. A., Lockley S. W., *Sleep, Health and Society: From Aetiology to Public Health*, Oxford University Press, Oxford, 2010.

Cappuccio F. P., Cooper D., D'Elia L., Strazzullo P., Miller M. A., 'Sleep duration predicts cardiovascular outcomes: a systematic review and meta-analysis of prospective studies', *Eur. Heart J.*, 32(12), June 2011, pp. 1,484–1,492.

Challamel M.-J., Thirion M., *Le Sommeil, le rêve et l'enfant*, Albin Michel, 1999.

Chang H.-M., Mai F.-D., Chen B.-J., Wu U.-I., Huang Y.-L., Lan C.-T., Ling Y.-C., 'Sleep deprivation predisposes liver to oxidative stress and phospholipid damage: a quantitative molecular imaging study', *J. Anat.*, 212(3), April 2008, pp. 295–305.

Chellappa S. L., Steiner R., Oelhafen P., Lang D., Götz T., Krebs J., Cajochen C., 'Acute exposure to evening blue-enriched light impacts on human sleep', *J. Sleep Res.*, 22(5), Oct. 2013, pp. 573–580. CIRC, 'Painting firefighting, and shiftwork', *IARC Monogr. Eval. Carcinog. Risks Hum.*, 98, 2010, pp. 9–764.

Cohen S., Doyle W. J., Alper C. M., Janicki-Deverts D., Turner R. B., 'Sleep habits and susceptibility to the common cold', *Arch. Intern. Med.*, 169(1), Jan. 2009, pp. 62–67.

Collis S. J. & Boulton S. J., 'Emerging links between the biological clock and the DNA damage response', *Chromosoma*, 116(4), Aug. 2007, pp. 331–339.

Cordi M. J., Schlarb A. A., Rasch B., 'Deepening sleep by hypnotic suggestion', *Sleep*, 37(6), June 2014, pp. 1,143–1,152.

Coren S., 'Sleep deprivation, psychosis and mental efficiency', *Psychiatric Times*, 15(3), March 1998.

Crary J., *24/7: Late Capitalism and the Ends of Sleep*, Verso Books, 2013, p. 127.

Cross N., Terpening Z., Rogers N. L., Duffy S. L., Hickie I. B., Lewis S. J., Naismith S. L., 'Napping in older people "at risk" of dementia: relationships with depression, cognition, medical burden and sleep quality', *J. Sleep Res.*, 24(5), Oct. 2015, pp. 494–502.

Dahl R. E., 'Biological, developmental, and neurobehavioral factors relevant to adolescent driving risks', *Am. J. Prev. Med.*, 35(3 suppl.), Sept. 2008, pp. 278–284.

Dalí S., *50 Secrets of Magic Craftsmanship*, tr. Haakon M. Chevalier, Dover Publications, Inc., New York, 1992, p. 35.

Debellemaniere E., Gomez-Merino D., Erblang M., Dorey R., Genot M., Perreaut-Pierre E., Pisani A., Rocco L., Sauvet F., Léger D., Rabat A., Chennaoui M., 'Using relaxation techniques to improve sleep during naps', *Ind. Health*, 56(3), June 2018, pp. 220–227.

Dejours C., *Souffrance en France. La banalisation de l'injustice sociale*, Seuil, 1998.

Dejours C., 'Activisme professionnel: masochisme, compulsivité ou aliénation?' [2004], in *Situations du travail*, PUF, 2016.

De Vivo L., Bellesi M., Marshall W., Bushong E. A., Ellisman M. H., Tononi G., Cirelli C., 'Ultrastructural evidence for synaptic scaling across the wake/sleep cycle', *Science*, 355(6324), Feb. 2017, pp. 507–510.

Dickinson D. L., Wolkow A. P., Rajaratnam S. M. W., Drummond S. P. A., 'Personal sleep debt and daytime sleepiness mediate the relationship between sleep and mental health outcomes in young adults', *Depress. Anxiety*, 35(8), Aug. 2018, pp. 775–783.

Diekelmann S., Landolt H. P., Lahl O., Born J., Wagner U., 'Sleep loss produces false memories', *PLoS One*, 3(10), 2008.

Diering G. H., Nirujogi R. S., Roth R. H., Worley P. F., Pandey A., Huganir R. L., 'Homer1a drives homeostatic scaling-down of excitatory synapses during sleep', *Science*, 355(6324), Feb. 2017, pp. 511–515.

Dijk D.-J., Archer S. N., 'PERIOD3, circadian phenotypes, and sleep homeostasis', *Sleep Medicine Reviews*, 14, 2010, pp. 151–160.

Dimitrov S., Benedict C., Heutling D., Westermann J., Born J., Lange T., 'Cortisol and epinephrine control opposing circadian rhythms in T cell subsets', *Blood*, 113(21), May 2009, pp. 5,134–5,143.

Dinges D. F., 'Adult napping and its effects on ability to function', in C. Stampi (ed.), *Why We Nap: Evolution, Chronobiology, and Functions of Polyphasic and Ultrashort Sleep*, Birkhäuser, Boston, MA, 1992, pp. 118–134.

Dinges D. F., Douglas S. D., Zaugg L., Campbell D. E., McMann J. M., Whitehouse W. G., Orne E. C., Kapoor S. C., Icaza E., Orne M. T., 'Leukocytosis and natural killer cell function parallel neurobehavioral fatigue induced by 64 hours of sleep deprivation', *J. Clin. Invest.*, 93(5), May 1994, pp. 1,930–1,939.

Dudley C. A., Erbel-Sieler C., Estill S. J., Reick M., Franken P., Pitts S., McKnight S. L., 'Altered patterns of sleep and behavioral adaptability in NPAS2-deficient mice', *Science*, 301(5631), July 2003, pp. 379–383.

Duffy J. F., Cain S. W., Chang A.-M., Phillips A. J. K., Münch M. Y., Gronfier C., Wyatt J. K., Dijk D. J., Wright K. P., Czeisler C. A., 'Sex difference in the near-24-hour intrinsic period of the human circadian timing system', *PNAS*, 108(3), 2011.

Edwards R. R., Almeida D. M., Klick B., Haythornthwaite J. A., Smith M. T., 'Duration of sleep contributes to next-day pain report in the general population', *Pain*, 137(1), July 2008, pp. 202–207.

Evans F. J., Cook M. R., Cohen H. D., Orne E. C., Orne M. T., 'Appetitive and replacement naps: EEG and behavior', *Science*, 197(4304), Aug. 1977, pp. 687–689.

Everson C. A., 'Sustained sleep deprivation impairs host defense', *Am. J. Physiol.*, 265(5/2), Nov. 1993, pp. 1,148–1,154.

Everson C. A., Laatsch C. D., Hogg N., 'Antioxidant defense responses to sleep loss and sleep recovery', *Am. J. Physiol. Regul. Integr. Comp. Physiol.*, 288(2), Feb. 2005, pp. 374–383.

Faraut B., Boudjeltia K. Z., Dyzma M., Rousseau A., David E., Stenuit P., Franck T., Van Antwerpen P., Vanhaeverbeek M., Kerkhofs M., 'Benefits of napping and an extended duration of recovery sleep on alertness and immune cells after acute sleep restriction', *Brain Behav. Immun.*, 25(1), Jan. 2011, pp. 16–24.

Faraut B., Touchette E., Gamble H., Royant-Parola S., Safar M. E., Varsat B., Léger D., 'Short sleep duration and increased risk of hypertension: a primary care medicine investigation', *J. Hypertens.*, 30(7), July 2012, pp. 1,354–1,363.

Faraut B., Bayon V., Léger D., 'Neuroendocrine, immune and oxidative stress in shift workers', *Sleep Med. Rev.*, 17(6), Dec. 2013, pp. 433–444.

Faraut B., Léger D., Medkour T., Dubois A., Bayon V., Chennaoui M., Perrot S., 'Napping reverses increased pain sensitivity due to sleep restriction', *PLoS One*, 20(2), Feb. 2015a.

Faraut B., Nakib S., Drogou C., Elbaz M., Sauvet F., De Bandt J. P., Léger D., 'Napping reverses the salivary interleukin-6 and urinary norepinephrine changes induced by sleep restriction', *J. Clin. Endocrinol. Metab.*, 100(3), March 2015b, pp. 416–426.

Faraut B., Andrillon T., Vecchierini M.-F., Léger D., 'Napping: a public health issue from epidemiological to laboratory studies', *Sleep Med. Rev.*, 35, Oct. 2017, pp. 85–100.

Faraut B., Léger D., Drogou C., Gauriau C., Dubois A., Servonnet A., Van Beers P., Guillard M., Gomez D., Andrillon T., Chennaoui M., 'Daytime exposure to blue-enriched light counters the effects of sleep restriction on cortisol, testosterone, alpha-amylase and executive processes', *Front. Neurosci.*, 13, Jan. 2020.

Fernandez G., Gatounes F., Herbain P., Vallejo P., *Nous, conducteurs de trains*, La Dispute, 2003.

Ferrie J. E., Shipley M. J., Cappuccio F. P., Brunner E., Miller M. A., Kumari M., Marmot M. G., 'A prospective study of change in sleep duration: associations with mortality in the Whitehall II cohort', *Sleep*, 30(12), Dec. 2007, pp. 1,659–1,666.

Ficca G., Axelsson J., Mollicone D. J., Muto V., Vitiello M. V., 'Naps, cognition and performance', *Sleep Med. Rev.*, 14(4), Aug. 2010, pp. 249–258.

Fischer S., Hallschmid M., Elsner A. L., Born J., 'Sleep forms memory for finger skills', *PNAS*, 99(18), Sept. 2002, pp. 11,987–11,991.

Franken P., Lopez-Molina L., Marcacci L., Schibler U., Tafti M., 'The transcription factor DBP affects circadian sleep consolidation and rhythmic EEG activity', *J. Neurosci.*, 20(2), Jan. 2000, pp. 617–625.

Franken P., Dudley C. A., Estill S. J., Barakat M., Thomason R., O'Hara B. F., McKnight S. L., 'NPAS2 as a transcriptional regulator of non-rapid eye movement sleep: genotype and sex interactions', *PNAS*, 103(18), May 2006, pp. 7,118–7,123.

Fröberg J. E., 'Twenty-four-hour patterns in human performance, subjective and physiological variables and differences between morning and evening active subjects', *Biol. Psychol.*, 5(2), June 1977, pp. 119–134.

Gais S., Albouy G., Boly M., Dang-Vu T. T., Darsaud A., Desseilles M., Rauchs G., Schabus M., Sterpenich V., Vandewalle G., Maquet P., Peigneux P., 'Sleep transforms the cerebral trace of declarative memories', *PNAS*, 104(47), Nov. 2007, pp. 18,778–18,783.

Gamble K. L., Motsinger-Reif A. A., Hida A., Borsetti H. M., Servick S. V., Ciarleglio C. M., Robbins S., Hicks J., Carver K., Hamilton N., Wells N., Summar M. L., McMahon D. G., Johnson C. H., 'Shift work in nurses: contribution of phenotypes and genotypes to adaptation', *PLoS One*, 6(4), April 2011.

Gangwisch J. E., Heymsfield S. B., Boden-Albala B., Buijs R. M., Kreier F., Pickering T. G., Rundle A. G., Zammit G. K., Malaspina D., 'Sleep duration as a risk factor for diabetes incidence in a large US sample', *Sleep*, 30(12), Dec. 2007, pp. 1,667–1,673.

Garbarino S., Mascialino B., Penco M. A., Squarcia S., De Carli F., Nobili L., Beelke M., Cuomo G., Ferrillo F., 'Professional shift-work drivers who adopt prophylactic naps can reduce the risk of car accidents during night work', *Sleep*, 27(7), Nov. 2004, pp. 1,295–1,302.

Garbarino S., Durando P., Guglielmi O., Dini G., Bersi F., Fornarino S., Toletone A., Chiorri C., Magnavita N., 'Sleep apnea, sleep debt and daytime sleepiness are independently associated with road accidents. A cross-sectional study on truck drivers', *PLoS One*, 11(11), Nov. 2016.

Garnier G., *L'Oubli des peines. Une histoire du sommeil (1700-1850)*, Rennes, Presses universitaires de Rennes, 2013.

Genzel L., Kroes M. C., Dresler M., Battaglia F. P., 'Light sleep versus slow wave sleep in memory consolidation: a question of global versus local processes?', *Trends Neurosci.*, 37(1), Jan. 2014, pp. 10–19.

Goichot B., Brandenberger G., Saini J., Wittersheim G., Follenius M., 'Nocturnal plasma thyrotropin variations are related to slow-wave sleep', *J. Sleep Res.*, 1(3), Sept. 1992, pp. 186–190.

Gottlieb D. J., Punjabi N. M., Newman A. B., Resnick H. E., Redline S., Baldwin C. M., Nieto F. J., 'Association of sleep time with diabetes mellitus and impaired glucose tolerance', *Arch. Intern. Med.*, 165(8), April 2005, pp. 863–867.

Granda T. G., Liu X.-H., Smaaland R., Cermakian N., Filipski E., Sassone-Corsi P., Lévi F., 'Circadian regulation of cell cycle and apoptosis proteins in mouse bone marrow and tumor', *FASEB J.*, 9(2), Feb. 2005, pp. 304–306.

Gronfier C., 'Hommes et femmes: à chacun son horloge …', Inserm press release, 2011: https://presse.inserm.fr/ hommes-et-femmes-a-chacun-son-horloge/13656.

Gronfier C., 'Horloge circadienne et fonctions non visuelles : rôle de la lumière chez l'Homme', *Biol. Aujourd'hui*, 208(4), 2014, pp. 261–267.

Gronfier C., Chapotot F., Weibel L., Jouny C., Piquard F., Brandenberger G., 'Pulsatile cortisol secretion and EEG delta waves are controlled by two independent but synchronized generators', *Am. J. Physiol.*, 275(1), July 1998, pp. 94–100.

Gujar N., McDonald S. A., Nishida M., Walker M. P., 'A role for REM sleep in recalibrating the sensitivity of the human brain to specific emotions', *Cereb. Cortex*, 21(1), Jan. 2011, pp. 115–123.

Gulia K. K., Kumar V. M., 'Sleep disorders in the elderly: a growing challenge', *Psychogeriatrics*, 18(3), May 2018, pp. 155–165.

Hamermesh D. S., Stancanelli E., 'Long workweeks and strange hours', *ILR Review*, Cornell University, ILR School, 68(5), Oct. 2015, pp. 1,007–1,018.

Harman S. M., Metter E. J., Tobin J. D., Pearson J., Blackman M. R., 'Longitudinal effects of aging on serum total and free testosterone levels in healthy men. Baltimore Longitudinal Study of Aging', *J. Clin. Endocrinol. Metab.*, 86(2), Feb. 2001, pp. 724–731.

Hayashi M., Masuda A., Hori T., 'The alerting effects of caffeine, bright light and face washing after a short daytime nap', *Clin. Neurophysiol.*, 114(12), Dec. 2003, pp. 2,268–2,278.

Hayashi M., Motoyoshi N., Hori T., 'Recuperative power of a short daytime nap with or without stage 2 sleep', *Sleep*, 28(7), July 2005, pp. 829–836.

He Y., Jones C. R., Fujiki N., Xu Y., Guo B., Holder J. L., Rossner M. J., Nishino S., Fu Y. H., 'The transcriptional repressor DEC2 regulates sleep length in mammals', *Science*, 325(5942), Aug. 2009, pp. 866–870.

Heslop P., Smith G. D., Metcalfe C., Macleod J., Hart C., 'Sleep duration and mortality: the effect of short or long sleep duration on cardiovascular and all-cause mortality in working men and women', *Sleep Med.*, 3(4), July 2002, pp. 305–314.

Hoddes E., Zarcone V., Smythe H., Phillips R., Dement W. C., 'Quantification of sleepiness: a new approach', *Psychophysiology*, 10(4), July 1973, pp. 431–436.

Horváth K., Myers K., Foster R., Plunkett K., 'Napping facilitates word learning in early lexical development', *J. Sleep Res.*, 24(5), Oct. 2015, pp. 503–509.

Hublin C., Kaprio J., Partinen M., Koskenvuo M., 'Insufficient sleep — a population-based study in adults', *Sleep*, 24(4), Feb. 2001.

INPES (Institut national de prévention et d'éducation pour la santé) 2005: http://inpes.santepubliquefrance.fr/Barometres/barome-tre-sante-2010/pdf/SH-depression.pdf.

INSV (Institut national du sommeil et de la vigilance), 'La journée du sommeil', 2004–2,018: https://institut-sommeil-vigilance.org/archives.

Irwin M., McClintick J., Fortner M., White J., Gillin J. C., 'Partial night deprivation reduces natural killer and cellular immune responses in humans', *FASEB J.*, 10, 1996, pp. 643–653.

Irwin M., Thompson J., Miller C., Gillin J. C., Ziegler M., 'Effects of sleep and sleep deprivation on catecholamine and interleukin-2 levels in humans: clinical implications', *J. Clin. Endocrinol. Metab.*, 84(6), June 1999, pp. 1,979–1,985.

Irwin M. R., Olmstead R., Carroll J. E., 'Sleep disturbance, sleep duration, and inflammation: a systematic review and meta-analysis of cohort studies and experimental sleep deprivation', *Biol. Psychiatry*, 80(1), July 2016, pp. 40–52.

Jaadane I., Villalpando Rodriguez G. E., Boulenguez P., Chahory S., Carré S., Savoldelli M., Jonet L., BeharCohen F., Martinsons C., Torriglia A., 'Effects of white light-emitting diode (LED) exposure on retinal pigment epithelium in vivo', *J. Cell. Mol. Med.*, 21(12), Dec. 2017, pp. 3,453–3,466.

Jakubowski K. P., Hall M. H., Lee L., Matthews K. A., 'Temporal relationships between napping and nocturnal sleep in healthy adolescents', *Behav. Sleep Med.*, 15(4), July–Aug. 2017, pp. 257–269.

Johns M. W., 'A new method for measuring day time sleepiness: The Epworth sleepiness scale' *Sleep*, 14(6), Dec. 1991.

Jouvet M., *Pourquoi rêvons-nous? Pourquoi dormons-nous? Où, quand, comment?*, Odile Jacob, 2000.

Jung C. M., Melanson E. L., Frydendall E .J., Perreault L., Eckel R. H., Wright K. P., 'Energy expenditure during sleep, sleep deprivation and sleep following sleep deprivation in adult humans', *J. Physiol.*, 589(1), Jan. 2011, pp. 235–244.

Kaida K., Takahashi M., Otsuka Y., 'A short nap and natural bright light exposure improve positive mood status', *Ind. Health*, 45(2), April 2007, pp. 301–308.

Kanabrocki E. L., Murray D., Hermida R. C., Scott G. S., Bremner W. F., Ryan M., 'Circadian variation in oxidative stress markers in healthy and type II diabetic men', *Chronobiol. Int.*, 19, March 2002, pp. 423–439.

Kang H., Feng X., Zhang B., Guo E., Wang L., Qian Z., Liu P., Wen X., Xu W., Li Y., Jiang C., Wu Z., Zhang H., Liu A., 'The siesta habit is associated with a decreased risk of rupture of intracranial aneurysms', *Front. Neurol.*, 8, Sept. 2017, p. 451.

Karacan I., Moore C. A., Hirshkowitz M., Sahmay S., Narter E. M., Tokat Y., Tuncel L., 'Uterine activity during sleep', *Sleep*, 9(3), 1986, pp. 393–398.

Kerkhofs M., *Le Sommeil de A à Zzzz*, Labor, 2000.

Kerkhofs M., Boudjeltia K. Z., Stenuit P., Brohée D., Cauchie P., Vanhaeverbeek M., 'Sleep restriction increases blood neutrophils, total cholesterol and low density lipoprotein cholesterol in postmenopausal women: a preliminary study', *Maturitas*, 56(2), Feb. 2007, pp. 212–215.

Kim H. M., Lee S. W., 'Beneficial effects of appropriate sleep duration on depressive symptoms and perceived stress severity in a healthy population in Korea', *Korean J. Fam. Med.*, 39(1), Jan. 2018, pp. 57–61.

Kimura A., Kishimoto T., 'IL-6: regulator of Treg/Th17 balance', *Eur. J. Immunol.*, 40(7), July 2010, pp. 1,830–1,835.

King C. R., Knutson K. L., Rathouz P. J., Sidney S., Liu K., Lauderdale D. S., 'Short sleep duration and incident coronary artery calcification', *JAMA*, 300(24), Dec. 2008, pp. 2,859–2,866.

Klauer S. G., Dingus T. A., Neale V. L., Sudweeks J. D., Ramsey D. J., 'The impact of driver inattention on near-crash/crash risk: an analysis using the 100-car naturalistic driving study data', National Highway Traffic Safety Administration, Washington, 2006.

Knutson K. L., Van Cauter E., Rathouz P. J., DeLeire T., Lauderdale D. S., 'Trends in the prevalence of short sleepers in the USA: 1975–2006', *Sleep*, 33(1), Jan. 2010, pp. 37–45.

Koch P., Montagner H., Soussignan R., 'Variations of behavioral and physiological variables in children attending kindergarten and primary school', *Chronobiology*

International, no. 4, Feb. 1987, pp. 525–535.

Komada Y., Asaoka S., Abe T., Matsuura N., Kagimura T., Shirakawa S., Inoue Y., 'Relationship between napping pattern and nocturnal sleep among Japanese nursery school children', *Sleep Med.*, 13(1), Jan. 2012, pp. 107–110.

Kripke D. F., Garkinfel L., Wingard D. L., Klauber M. R., Marler M. R., 'Mortality associated with sleep duration and insomnia', *Arch. Gen. Psychiatry*, 59(2), Feb. 2002, pp. 131–136.

Kurdziel L., Duclos K., Spencer R. M. C., 'Sleep spindles in midday naps enhance learning in preschool children', *PNAS*, 110(43), Oct. 2013, pp. 17,267–17,272.

Lahl O., Wispel C., Willigens B., Pietrowsky R., 'An ultra short episode of sleep is sufficient to promote declarative memory performance', *J. Sleep Res.*, 17(1), March 2008, pp. 3–10.

Lampl M., Johnson M. L., 'Infant growth in length follows prolonged sleep and increased naps', *Sleep*, 34(5), May 2011, pp. 641–650.

Lange T., Perras B., Fehm H. L., Born J., 'Sleep enhances the human antibody response to hepatitis A vaccination', *Psychosom. Med.*, 65(5), Sept.–Oct. 2003, pp. 831–835.

Lange T., Dimitrov S., Born J., 'Effects of sleep and circadian rhythm on the human immune system', *Ann. N. Y. Acad. Sci.*, 1193(1), April 2010, pp. 48–59.

Lange T., Born J., 'The immune recovery function of sleep — tracked by neutrophil counts', *Brain Behav. Immun.*, 25(1), Jan. 2011, pp. 14–15.

Lasselin J., Rehman J. U., Åkerstedt T., Lekander M., Axelsson J., 'Effect of long-term sleep restriction and subsequent recovery sleep on the diurnal rhythms of white blood cell subpopulations', *Brain Behav. Immun.*, 47, July 2015, pp. 93–99.

Lavie P., 'Ultradian rhythms in arousal — the problem of masking', *Chronobiol. Int.*, 6(1), 1989, pp. 21–28.

Lavie P., Weler B., 'Timing of naps: effects on post-nap sleepiness levels', *Electroencephalogr. Clin. Neurophysiol.*, 72(3), March 1989, pp. 218–224.

Lee S., McCann D., Messenger J. C., *Working Time Around the World: Trends in Working Hours, Laws, and Policies in a Global Comparative Perspective*, Routledge, London, New York, 2007.

Léger D., *Les Troubles du sommeil*, PUF, 'Que sais-je?' 2017.

Léger D., Roscoat E. du, Bayon V., Guignard R., Pâquereau J., Beck F., 'Short sleep in young adults: insomnia or sleep debt? Prevalence and clinical description of short sleep in a representative sample of 1004 young adults from France', *Sleep Medicine*, 12(5), May 2011, pp. 454–462.

Léger D., Beck F., Richard J.-B., Godeau E., 'Total sleep time severely drops during adolescence', *PLoS ONE*, 7(10), 2012.

Léger D., Beck F., Richard J.-B., Sauvet F., Faraut B., 'The risks of sleeping "too much". Survey of a national representative sample of 24,671 adults (INPES health barometer)', *PLoS ONE*, 9(9), 2014.

Leproult R., Van Cauter E., 'Effect of 1 week of sleep restriction on testosterone levels in young healthy men', *JAMA*, 305(21), June 2011, pp. 2,173–2,174.

Lin J. F., Li F. D., Chen X. G., He F., Zhai Y. J., Pan X. Q., Wang X. Y., Zhang T., Yu M., 'Association of postlunch napping duration and night-time sleep duration with cognitive impairment in Chinese elderly: a cross-sectional study', *BMJ Open*, 8(12), Dec. 2018.

Lockley S. W., Evans E. E., Scheer F. A., Brainard G. C., Czeisler C. A., Aeschbach D., 'Short-wavelength sensitivity for the direct effects of light on alertness, vigilance, and the waking electroencephalogram in humans', *Sleep*, 29(2), Feb. 2006, pp. 161–168.

Loimaala A., Rontu R., Vuori I., Mercuri M., Lehtimäki T., Nenonen A., Bond M. G., 'Blood leukocyte count is a risk factor for intima-media thickening and subclinical carotid atherosclerosis in middle-aged men', *Atherosclerosis*, 188(2), Oct. 2006, pp. 363–369.

Lopez-Minguez J., Morosoli J. J., Madrid J. A., Garaulet M., Ordoñana J. R., 'Heritability of siesta and night-time sleep as continuously assessed by a circadian-related integrated measure', *Sci. Rep.*, 7(1), Sept. 2017, p. 12,340.

Lowden A., Holmback U., Åkerstedt T., Forslund J., Lennernas M., Forslund A., 'Performance and sleepiness during a 24h wake in constant conditions are affected by diet', *Biol. Psychol.*, 65(3), Feb. 2004, pp. 251–263.

Luboshitzky R., Herer P., Levi M., Shen-Orr Z., Lavie P., 'Relationship between rapid eye movement sleep and testosterone secretion in normal men', *J. Androl.*, 20(6), Nov.–Dec. 1999, pp. 731–737.

Luboshitzky R., Zabari Z., Shen-Orr Z., Herer P., Lavie P., 'Disruption of the nocturnal testosterone rhythm by sleep fragmentation in normal men', *J. Clin. Endocrinol. Metab.*, 86(3), March 2001, pp. 1,134–1,139.

Luboshitzky R., Shen-Orr Z., Herer P., 'Middle-aged men secrete less testosterone at night than young healthy men', *J. Clin. Endocrinol. Metab.*, 88(7), July 2003, pp. 3,160–3,166.

Lusardi P., Mugellini A., Preti P., Zoppi A., Derosa G., Fogari R., 'Effects of a restricted sleep regimen on ambulatory blood pressure monitoring in normotensive subjects', *Am. J. Hypertens.*, 9(5), May 1996, pp. 503–505.

Mantua J., Spencer R. M. C., 'The interactive effects of nocturnal sleep and daytime naps in relation to serum C-reactive protein', *Sleep Med.*, 16(10), Oct. 2015, pp. 1,213–1,216.

Mantua J., Spencer R. M. C., 'Exploring the nap paradox: are mid-day sleep bouts a friend or foe?', *Sleep Med.*, 37, Sept. 2017, pp. 88–97.

Markwald R. R., Melanson E. L., Smith M. R., Higgins J., Perreault L., Eckel R. H., Wright K. P. Jr., 'Impact of insufficient sleep on total daily energy expenditure, food intake, and weight gain', *PNAS*, 110(14), April 2013, pp. 5,695–5,700.

Mednick S., Nakayama K., Stickgold R., 'Sleep-dependent learning: a nap is as good as a night', *Nat. Neurosci.*, 6(7), July 2003, pp. 697–698.

Meerlo P., Havekes R., Steiger A., 'Chronically restricted or disrupted sleep as a causal factor in the development of depression', *Curr. Top Behav. Neurosci.*, 25, 2015, pp. 459–481.

Milner C. E., Fogel S. M., Cote K. A., 'Habitual napping moderates motor performance improvements following a short daytime nap', *Biol. Psychol.*, 73(2), Aug. 2006, pp. 141–156.

Milner C. E., Cote K. A., 'Benefits of napping in healthy adults: impact of nap length, time of day, age, and experience with napping', *J. Sleep Res.*, 18(2), June 2009, pp. 272–281.

Mindell J. A., Sadeh A., Kwon R., Goh D. Y., 'Cross-cultural differences in the sleep of preschool children', *Sleep Med.*, 14(12), Dec. 2013, pp. 1,283–1,289.

Montagner H., 'Les rythmes majeurs de l'enfant', *Informations sociales*, no. 153, March 2009, pp. 14–20.

Morita Y., Ogawa K., Uchida S., 'Napping after complex motor learning enhances juggling performance', *Sleep Sci.*, 9(2), April 2016, pp. 112–116.

Mumford L., *Technics and Civilization*, Routledge, London, 1934, pp. 14–15.

Naitoh P., 'Circadian cycles and restorative power of naps', in Johnson L. C., Tepas D. I., Colquhoun W. P., Colligan M. J. (ed.), *The Twenty-Four Hour Workday: Proceedings of a Symposium on Variations in Work-Sleep Schedules*, Cincinnati, OH, US Department of Health and Human Services, Public Health Service, Centers for Disease Control, National Institute for Occupational Safety and Health, DHHS (NIOSH) Publication, 81–127, July 1981, pp. 693–720.

Naska A., Oikonomou E., Trichopoulou A., Psaltopoulou T., Trichopoulos D., 'Siesta in healthy adults and coronary mortality in the general population', *Arch. Intern. Med.*, 167(3), Feb. 2007, pp. 296–301.

Naylor E., Bergmann B. M., Krauski K., Zee P. C., Takahashi J. S., Vitaterna M.

H., Turek F. W., 'The circadian clock mutation alters sleep homeostasis in the mouse', *J. Neurosci.*, 20(21), Nov. 2000, pp. 8,138–8,143.

NCSDR (National Center on Sleep Disorders Research) / NHTSA (National Highway Transportation Safety Administration), Expert Panel on Driver Fatigue and Sleepiness, 'Drowsy Driving and Automobile Crashes: Report and Recommendations', Washington, 1997: https://www.nhtsa.gov/sites/nhtsa.dot.gov/files/808707.pdf.

Nedeltcheva A. V., Kessler L., Imperial J., Penev P. D., 'Exposure to recurrent sleep restriction in the setting of high caloric intake and physical inactivity results in increased insulin resistance and reduced glucose tolerance', *J. Clin. Endocrinol. Metab.*, 94(9), Sept. 2009, pp. 3,242–3,250.

Nishida M., Walker M. P., 'Daytime naps, motor memory consolidation and regionally specific sleep spindles', *PLoS One*, 2(4), 2007, pp. 341.

NSF, 2009: https://sleepfoundation.org/sleep-polls-data/sleep-in-america-poll/2009-health-and-safety. 2013: https://sleepfoundation.org/sites/default/files/RPT495a.pdf. 2015: https://sleepfoundation.org/sleep-polls-data/2015-sleep-and-pain.

Ohayon M. M., Carskadon M. A., Guilleminault C., Vitiello M. V., 'Meta-analysis of quantitative sleep parameters from childhood to old age in healthy individuals: developing normative sleep values across the human lifespan', *Sleep*, 27(7), Nov. 2004, pp. 1,255–1,273.

Ohayon M. M., Milesi C., 'Artificial outdoor nighttime lights associate with altered sleep behavior in the American general population', *Sleep*, 39(6), June 2016, pp. 1,311–1,320.

Onen S. H., Alloui A., Gross A., Eschallier A., Dubray C., 'The effects of total sleep deprivation, selective sleep interruption and sleep recovery on pain tolerance thresholds in healthy subjects', *J. Sleep Res.*, 10(1), March 2001, pp. 35–42.

Paquot T., *L'Art de la sieste*, Zulma, 2008.

Patel S. R., Malhotra A., Gao X., Hu F. B., Neuman M. I., Fawzi W. W., 'A prospective study of sleep duration and pneumonia risk in women', *Sleep*, 35(1), Jan. 2012, pp. 97–101.

Peigneux P., Laureys S., Fuchs S., Collette F., Perrin F., Reggers J., Phillips C., Degueldre C., Del Fiore G., Aerts J., Luxen A., Maquet P., 'Are spatial memories strengthened in the human hippocampus during slow wave sleep?', *Neuron.*, 44(3), Oct. 2004, pp. 535–545.

Pellegrino R., Kavakli I. H., Goel N., Cardinale C. J., Dinges D. F., Kuna S. T.,

Maislin G., Van Dongen H. P. A., Tufik S., Hogenesch J. B., Hakonarson H., Pack A. I., 'A novel BHLHE41 variant is associated with short sleep and resistance to sleep deprivation in humans', *Sleep*, 37(8), Aug. 2014, pp. 1,327–1,336.

Periasamy S., Hsu D.-Z., Fu Y.-H., Liu M.-Y., 'Sleep deprivation-induced multi-organ injury: role of oxidative stress and inflammation', *EXCLI Journal*, 14, 2015, pp. 672–683.

Picarsic J. L., Glynn N. W., Taylor C. A., Katula J. A., Goldman S. E., Studenski S. A., Newman A. B., 'Self-reported napping and duration and quality of sleep in the lifestyle interventions and independence for elders pilot study', *J. Am. Geriatr. Soc.*, 56(9), Sept. 2008, pp. 1,674–1,680.

Plihal W., Born J., 'Effects of early and late nocturnal sleep on declarative and procedural memory', *J. Cogn. Neurosci.*, 9(4), July 1997, pp. 534–547.

Prather A. A., Janicki-Deverts D., Hall M. H., Cohen S., 'Behaviorally assessed sleep and susceptibility to the common cold', *Sleep*, 38(9), Sept. 2015, pp. 1,353–1,359.

Ramachandruni S., Handberg E., Sheps D. S., 'Acute and chronic psychological stress in coronary disease', *Curr. Opin. Cardiol.*, 19(5), Sept. 2004, pp. 494–499.

Raskind M. A., Peskind E. R., Kanter E. D., Petrie E. C., Radant A., Thompson C. E., Dobie D. J., Hoff D., Rein R. J., Straits-Tröster K., Thomas R. G., McFall M. M., 'Reduction of nightmares and other PTSD symptoms in combat veterans by prazosin: a placebo-controlled study', *Am. J. Psychiatry*, 160(2), Feb. 2003, pp. 371–373.

Revell V. L., Arendt J., Fogg L. F., Skene D. J., 'Alerting effects of light are sensitive to very short wavelengths', *Neurosci. Lett.*, 399(1–2), May 2006, pp. 96–100.

Roach G. D., Matthews R., Naweed A., Kontou T. G., Sargent C., 'Flat-out napping: the quantity and quality of sleep obtained in a seat during the daytime increase as the angle of recline of the seat increases', *Chronobiol. Int.*, 35(6), June 2018, pp. 872–883.

Roehrs T. A., Harris E., Randall S., Roth T., 'Pain sensitivity and recovery from mild chronic sleep loss', *Sleep*, 35(12), Dec. 2012, pp. 1,667–1,672.

Roenneberg T., Kuehnle T., Pramstaller P. P., Ricken J., Havel M., Guth A., Merrow M., 'A marker for the end of adolescence', *Curr. Biol.*, 14(24), Dec. 2004, pp. 1,038–1,039.

Roenneberg T., Kuehnle T., Juda M., Kantermann T., Allebrandt K., Gordijn M., Merrow M., 'Epidemiology of the human circadian clock', *Sleep Med. Rev.*, 11(6), Dec. 2007, pp. 429–438.

Rowshan Ravan A., Bengtsson C., Lissner L., Lapidus L., Björkelund C.,

'Thirty-six-year secular trends in sleep duration and sleep satisfaction, and associations with mental stress and socioeconomic factors — results of the Population Study of Women in Gothenburg, Sweden', *J. Sleep Res.*, 19(3), Sept. 2010, pp. 496–503.

Ruggiero C., Metter E. J., Cherubini A., Maggio M., Sen R., Najjar S. S., Windham G. B., Ble A., Senin U., Ferrucci L., 'White blood cell count and mortality in the Baltimore Longitudinal Study of Aging', *J. Am. Coll. Cardiol.*, 49(18), May 2007, pp. 1,841–1,850.

Ryu S. Y, Kim K. S., Han M. A., 'Factors associated with sleep duration in Korean adults: results of a 2008 community health survey in Gwangju Metropolitan City, Korea', *J. Korean Med. Sci.*, 26(9), Sept. 2011, pp. 1,124–1,131.

Sagaspe P., Taillard J., Chaumet G., Moore N., Bioulac B., Philip P., 'Aging and nocturnal driving: better with coffee or a nap? A randomized study', *Sleep*, 30(12), Dec. 2007, pp. 1,808–1,813.

Sassin J. F., Parker D. C., Mace J. W., Gotlin R. W., Johnson L. C., Rossman L. G., 'Human growth hormone release: relation to slow-wave sleep and sleep-waking cycles', *Science*, 165(3892), Aug. 1969, pp. 513–515.

Sayón-Orea C., Bes-Rastrollo M., Carlos S., Beunza J. J., Basterra-Gortari F. J., Martínez-González M. A., 'Association between sleeping hours and siesta and the risk of obesity: the SUN Mediterranean Cohort', *Obes. Facts*, 6(4), 2013, pp. 337–347.

Scheer F. A. J. L., Hilton M. F., Mantzoros C. S., Shea S. A., 'Adverse metabolic and cardiovascular consequences of circadian misalignment', *PNAS*, 106(11), March 2009, pp. 4,453–4,458.

Schibler U., Sassone-Corsi P., 'A web of circadian pacemakers', *Cell*, 111(7), Dec. 2002, pp. 919–922.

Schmid S. M., Hallschmid M., Jauch-Chara K., Wilms B., Lehnert H., Born J., Schultes B., 'Disturbed glucoregulatory response to food intake after moderate sleep restriction', *Sleep*, 34(3), March 2011, pp. 371-377.

Schweitzer P. K., Randazzo A. C., Stone K., Erman M., Walsh J. K., 'Laboratory and field studies of naps and caffeine as practical countermeasures for sleep-wake problems associated with night work', *Sleep*, 29(1), Jan. 2006, pp. 39–50.

Schwob M., *Les Rythmes du corps. Chronobiologie de l'alimentation, du sommeil, de la santé* ..., Odile Jacob, 2007.

Seehagen S., Konrad C., Herbert J. S., Schneider S., 'Timely sleep facilitates declarative memory consolidation in infants', PNAS, 112(5), Feb. 2015, pp. 1,625–1,629.

SFMT (Société française de médecine du travail), 'Surveillance médico-professionnelle des travailleurs postés et/ou de nuit', June 2012: http://www.chu-rouen.fr/sfmt/pages/accueil.php.

Simon C., Gronfier C., Schlienger J. L., Brandenberger G., 'Circadian and ultradian variations of leptin in normal man under continuous enteral nutrition: relationship to sleep and body temperature', *J. Clin. Endocrinol. Metab.*, 83(6), June 1998, pp. 1,893–1,899.

Spiegel K., Follenius M., Simon C., Saini J., Ehrhart J., Brandenberger G., 'Prolactin secretion and sleep', *Sleep*, 17(1), Feb. 1994, pp. 20–27.

Spiegel K., Luthringer R., Follenius M., Schaltenbrand N., Macher J. P., Muzet A., Brandenberger G., 'Temporal relationship between prolactin secretion and slow-wave electroencephalic activity during sleep', *Sleep*, 8(7), Sept. 1995, pp. 543–548.

Spiegel K., Leproult R., Van Cauter E., 'Impact of sleep debt on metabolic and endocrine function', *Lancet*, 354(9188), Oct. 1999, pp. 1,435–1,439.

Spiegel K., Sheridan J. F., Van Cauter E., 'Effect of sleep deprivation on response to immunization', *JAMA*, 288(12), Sept. 2002, pp. 1,471–1,472.

Spiegel K., Tasali E., Penev P., Van Cauter E., 'Brief communication: sleep curtailment in healthy young men is associated with decreased leptin levels, elevated ghrelin levels, and increased hunger and appetite', *Ann. Intern. Med.*, 141(11), Dec. 2004, pp. 846–850.

Spiegel K., Tasali E., Leproult R., Scherberg N., Van Cauter E., 'Twenty-four-hour profiles of acylated and total ghrelin: relationship with glucose levels and impact of time of day and sleep', *J. Clin. Endocrinol. Metab.*, 96(2), Feb. 2011, pp. 486–493.

Squire L. R., Kandel E. R. [1999], *La Mémoire. De l'esprit aux molécules*, Flammarion, Coll. 'Champs', 2005.

Stickgold R., Hobson J. A., Fosse R., Fosse M., 'Sleep, learning, and dreams: off-line memory reprocessing', *Science*, 294(5544), Nov. 2001, pp. 1,052–1,057.

Straif K., Baan R., Grosse Y., Secretan B., El Ghissassi F., Bouvard V., Altieri A., Benbrahim-Tallaa L., Cogliano V., 'Carcinogenicity of shift-work, painting, and fire-fighting', *Lancet Oncol.*, 8(12), Dec. 2007, pp. 1,065–1,066.

Suzuki H., Uchiyama M., Tagaya H., Ozaki A., Kuriyama K., Aritake S., Shibui K., Tan X., Kamei Y., Kuga R., 'Dreaming during non-rapid eye movement sleep in the absence of prior rapid eye movement sleep', *Sleep*, 27(8), Dec. 2004, pp. 1,486–1,490.

Takahashi M., Arito H., 'Maintenance of alertness and performance by a brief nap after lunch under prior sleep deficit', *Sleep*, 23(6), Sept. 2000, pp. 813–819.

Takahashi M., Iwakiri K., Sotoyama M., Hirata M., Hisanaga N., 'Musculoskeletal pain and night-shift naps in nursing home care workers', *Occup. Med.*, 59(3), May 2009, pp. 197–200.

Takahashi Y., Kipnis D. M., Daughaday W. H., 'Growth hormone secretion during sleep', *J. Clin. Invest.*, 47(9), Sept. 1968, pp. 2,079–2,090.

Theadom A., Cropley M., Kantermann T., 'Daytime napping associated with increased symptom severity in fibromyalgia syndrome', *BMC Musculoskelet. Disord.*, 16(13), Feb. 2015.

Thibault M., *Ouvriers malgré tout. Enquête sur les ateliers de maintenance des trains de la Régie autonome des transports parisiens*, éditions Raisons d'agir, Paris, 2013.

Thibault M., 'Métro, boulot, chrono', *Le Monde diplomatique*, Nov. 2014, pp. 19.

Thorleifsdottir B., Björnsson J. K., Benediktsdottir B., Gislason T., Kristbjarnarson H., 'Sleep and sleep habits from childhood to young adulthood over a 10-year period', *J. Psychosom. Res.*, 53(1), July 2002, pp. 529–537.

Tiede W., Magerl W., Baumgärtner U., Durrer B., Ehlert U., Treede R. D., 'Sleep restriction attenuates amplitudes and attentional modulation of pain-related evoked potentials, but augments pain ratings in healthy volunteers', *Pain*, 148(1), Jan. 2010, pp. 36–42.

Tietzel A. J., Lack L. C., 'The recuperative value of brief and ultrabrief naps on alertness and cognitive performance', *J. Sleep Res.*, 11(3), Sept. 2002, pp. 213–218.

Touchette E., Petit D., Seguin J. R., Boivin M., Tremblay R. E., Montplaisir J. Y., 'Associations between sleep duration patterns and behavioral/cognitive functioning at school entry', *Sleep*, 30(9), Sept. 2007, pp. 1,213–1,219.

Touchette E., Dionne G., Forget-Dubois N., Petit D., Pérusse D., Falissard B., Tremblay R. E., Boivin M., Montplaisir J. Y., 'Genetic and environmental influences on daytime and night-time sleep duration in early childhood', *Pediatrics*, 131(6), June 2013, pp. 1,874–1,880.

Trichopoulos D., Tzonou A., Christopoulos C., Havatzoglou S., Trichopoulou A., 'Does a siesta protect from coronary heart disease?', *Lancet*, 2(8553), Aug. 1987, pp. 269–270.

Tsapanou A., Gu Y., O'Shea D., Eich T., Tang M. X., Schupf N., Manly J., Zimmerman M., Scarmeas N., Stern Y., 'Daytime somnolence as an early sign of cognitive decline in a community-based study of older people', *Int. J. Geriatr. Psychiatry*, 31(3), March 2016, pp. 247–255.

Vallat R., Lajnef T., Eichenlaub J. B., Berthomier C., Jerbi K., Morlet D., Ruby P. M., 'Increased evoked potentials to arousing auditory stimuli during sleep:

implication for the understanding of dream recall', *Front. Hum. Neurosci.*, 11, March 2017, art. 132.

Van Cauter E., Caufriez A., Kerkhofs M., Van Onderbergen A., Thorner M. O., Copinschi G., 'Sleep, awakenings, and insulin-like growth factor-I modulate the growth hormone (GH) secretory response to GH-releasing hormone', *J. Clin. Endocrinol. Metab.*, 74(6), June 1992, pp. 1,451–1,459.

Van Cauter E., Copinschi G., 'Interrelationships between growth hormone and sleep', *Growth Hormone & IGF Research*, 10, April 2000, pp. 57–62.

Van Cauter E., Latta F., Nedeltcheva A., Spiegel K., Leproult R., Vandenbril C., Weiss R., Mockel J., Legros J.-J., Copinschi G., 'Reciprocal interactions between the GH axis and sleep', *Growth Hormone & IGF Research*, 14, June 2004, pp. 10–17.

Van der Helm E., Gujar N., Walker M. P., 'Sleep deprivation impairs the accurate recognition of human emotions', *Sleep*, 33(3), March 2010, pp. 335–342.

Vela-Bueno A., Fernandez-Mendoza J., Olavarrieta-Bernardino S., Vgontzas A. N., Bixler E. O., Cruz-Troca J. J. de la, Rodriguez-Muñoz A., Olivan-Palacios J., 'Sleep and behavioral correlates of napping among young adults: a survey of first-year university students in Madrid, Spain', *J. Am. Coll. Health*, 57(2), Sept.–Oct. 2008, pp. 150–158.

Vezoli J., Fifel K., Leviel V., Dehay C., Kennedy H., Cooper H. M., Gronfier C., Procyk E., 'Early presymptomatic and long-term changes of rest activity cycles and cognitive behavior in a MPTP-monkey model of Parkinson's disease', *PLoS One*, 6(8), 2011.

Vgontzas A. N., Pejovic S., Zoumakis E., Lin H. M., Bixler E. O., Basta M., Fang J., Sarrigiannidis A., Chrousos G. P., 'Daytime napping after a night of sleep loss decreases sleepiness, improves performance, and causes beneficial changes in cortisol and interleukin-6 secretion', *Am. J. Physiol. Endocrinol. Metab.*, 292(1), Jan. 2007, pp. 253–261.

Vincent C., 'Christophe Dejours, psychiatre: "Les soignants sont contraints d'apporter leur concours à des actes qu'ils réprouvent"', *Le Monde*, 'Idées', 15 Feb. 2018.

Vogel G. W., Thurmond A., Gibbons P., Sloan K., Walker M. P., 'REM sleep reduction effects on depression syndromes', *Arch. Gen. Psychiatry*, 32(6), June 1975, pp. 765–777.

Wagner U., Gais S., Born J., 'Emotional memory formation is enhanced across sleep intervals with high amounts of rapid eye movement sleep', *Learn Mem.*,

8(2), March–April 2001, pp. 112–119.

Wagner U., Degirmenci M., Drosopoulos S., Perras B., Born J., 'Effects of cortisol suppression on sleep-associated consolidation of neutral and emotional memory', *Biol. Psychiatry*, 58(11), Dec. 2005, pp. 885–893.

Walker M., *Pourquoi nous dormons. Le pouvoir du sommeil et des rêves*, La Découverte, 2018.

Walker M. P., Brakefield T., Morgan A., Hobson J. A., Stickgold R., 'Practice with sleep makes perfect: sleep-dependent motor skill learning', *Neuron.*, 35(1), July 2002, pp. 205–211.

Weissbluth M., 'Naps in children: 6 months–7 years', *Sleep*, 18(2), Feb. 1995, pp. 82–87.

West K. E., Jablonski M. R., Warfield B., Cecil K. S., James M., Ayers M. A., Maida J., Bowen C., Sliney D. H., Rollag M. D., Hanifin J. P., Brainard G. C., 'Blue light from light-emitting diodes elicits a dose-dependent suppression of melatonin in humans', *J. Appl. Physiol.* [1985], 110(3), March 2011, pp. 619–626.

Wiegand M., Riemann D., Schreiber W., Lauer C. J., Berger M., 'Effect of morning and afternoon naps on mood after total sleep deprivation in patients with major depression', *Biol. Psychiatry*, 33(6), March 1993, pp. 467–476.

Wilson M. A., McNaughton B. L., 'Reactivation of hippocampal ensemble memories during sleep', *Science*, 265(5172), July 1994, pp. 676–679.

Wren A. M., Seal L. J., Cohen M. A., Brynes A. E., Frost G. S., Murphy K. G., Dhillo W. S., Ghatei M. A., Bloom S. R., 'Ghrelin enhances appetite and increases food intake in humans', *J. Clin. Endocrinol. Metab.*, 86(12), Dec. 2011, pp. 5,992–5,995.

Xie L., Kang H., Xu Q., Chen M. J., Liao Y., Thiyagarajan M., O'Donnell J., Christensen D. J., Nicholson C., Iliff J. J., Takano T., Deane R., Nedergaard M., 'Sleep drives metabolite clearance from the adult brain', *Science*, 342(6156), Oct. 2013, pp. 373–377.

Yamada T., Hara K., Shojima N., Yamauchi T., Kadowaki T., 'Daytime napping and the risk of cardiovascular disease and all-cause mortality: a prospective study and dose-response meta-analysis', *Sleep*, 38(12), Dec. 2015, pp. 1,945–1,953.

Yoon I.-Y., Kripke D. F., Youngstedt S. D., Elliott J. A., 'Actigraphy suggests age-related differences in napping and nocturnal sleep', *J. Sleep Res.*, 12(2), June 2003, pp. 87–93.

Zaregarizi M., Edwards B., George K., Harrison Y., Jones H., Atkinson G., 'Acute changes in cardiovascular function during the onset period of daytime sleep:

comparison to lying awake and standing', *J. Appl. Physiol.* [1985], 103(4), Oct. 2007, pp. 1,332–1,338.

Zhai L., Zhang H., Zhang D., 'Sleep duration and depression among adults: a meta-analysis of prospective studies', *Depress. Anxiety*, 32(9), Sept. 2015, pp. 664–670.

Zhao D., Zhang Q., Fu M., Tang Y., Zhao Y., 'Effects of physical positions on sleep architectures and post-nap functions among habitual nappers', *Biol. Psychol.*, 83(3), March 2010, pp. 207–213.

Zhong X., Xiao Y., Huang R., Huang X. Z., 'The effects of overnight sleep deprivation on cardiovascular autonomic modulation', *Zhonghua Nei Ke Za Zhi*, 44(8), Aug. 2005, pp. 577–580.

Zhou Y. Q., Liu Z., Liu Z.-H., Chen S.-P., Li M., Shahveranov A., Ye D.-W., Tian Y.-K., 'Interleukin-6: an emerging regulator of pathological pain', *J. Neuroinflammation*, 13(1), June 2016, pp. 141.

Notes

Introduction

1   Beck et al. 2012; ANSES 2018.
2   Garnier 2013, pp. 59, 40, and 108.
3   Coren 1998.
4   Ibid.

1. The Ins and Outs of Sleep

1   These nuclei are located just above the optic chiasm — where the two optic nerves converge — which reports the current level of daylight intensity to the master clock, enabling the clock to resynchronise with it every day.
2   Gronfier 2014.
3   Granda et al. 2005 (for cell division); Collis & Boulton 2007 (for DNA repair).
4   This experiment was first conducted in 1729 by the French astronomer Jean-Jacques Dortous de Mairan (Jouvet 2000, p. 45).
5   Duffy et al. 2011

6   Roenneberg et al. 2007. The study points out that this disparity is only proven till menopause.
7   Roenneberg et al. 2004. Age and sex are not the only determining factors: the alleles of one of the genes involved in adjusting the period of our biological clock might also have an influence on the nature of our chronotype (Dijk & Archer 2010).
8   Gronfier 2014.
9   Ibid., p. 48.
10 Chellappa et al. 2013.
11 Gronfier 2014.
12 http://www.non-24.com.
13 Vezoli et al. 2011.
14 Straif et al. 2007.
15 Schwob 2007, p. 203.
16 Ibid., p. 173.
17 Schibler & Sassone-Corsi 2002.
18 Adrien 2007, pp. 81–86.
19 Garnier 2013, p. 153.
20 This term was coined by the

great French neurobiologist Michel Jouvet (1925–2017).

21 Karacan et al. 1986.

22 Kerkhofs 2000, p. 14.

23 Faraut et al. 2017.

24 Challamel & Thirion 1999, p. 8.

25 Léger et al. 2012.

26 It decreases fairly soon (between six and ten years of age) to the 20–25 per cent observed in adulthood.

27 Challamel & Thirion 1999, p. 87.

28 Corresponding to a gradual decline in deep slow-wave sleep of 10–15 per cent between the ages of five and 15 (Ohayon et al. 2004).

29 German chronobiologist Till Roenneberg went about verifying this estimate in one of the above-mentioned studies involving 55,000 respondents aged 16–86: to partially offset sleep deprivation on working days, he averaged the figures for hours of sleep during periods of work (weeknights) and holidays (weekends) (Roenneberg et al. 2007).

30 Kripke et al. 2002. To explain why, one would have to study a large population of short and long sleepers who remain in perfect health over the course of several decades — which would be an extremely complicated undertaking.

31 Kerkhofs 2000, p. 30.

32 Léger et al. 2014.

33 Basner et al. 2014.

34 Aeschbach et al. 2003.

35 Aeschbach et al. 2001.

36 Cappuccio et al. 2010, p. 269.

37 This subjectivity is all the more pronounced among men, who tend to underestimate their actual hours of sleep — probably to vaunt their stamina.

38 Allebrandt et al. 2013.

39 Billiard 2000, p. 67.

40 Dijk & Archer 2010.

41 Naylor et al. 2000; Franken et al. 2000 and 2006; Dudley et al. 2003.

42 He et al. 2009.

43 Pellegrino et al. 2014.

2. Healing Sleep

1 Women are never picked for this type of experiment because, as noted above, the menstrual cycle can significantly affect the endocrine system.

2 Takahashi et al. 1968; Sassin

et al. 1969.

3   Van Cauter et al. 1992; Van Cauter & Copinschi 2000.

4   Van Cauter et al. 2004.

5   Lange et al. 2010.

6   Prolactin also plays a key part in reproduction and lactation. So a prolonged lack of sleep leading to a prolactin deficiency might affect a woman's fertility.

7   Spiegel et al. 1994 and 1995.

8   Faraut et al. 2011.

9   Axelsson et al. 2005

10 Luboshitzky et al. 1999.

11 Axelsson et al. 2005.

12 Leproult & Van Cauter 2011.

13 Luboshitzky et al. 2003.

14 Harman et al. 2001

15 Leproult & Van Cauter 2011 (for limited sleep); Luboshitzky et al. 2001 (for fragmented sleep); Axelsson et al. 2005 (for a sleepless night).

16 Bremner 2010.

17 Simon et al. 1998.

18 Wren 2001 and Spiegel 2011.

19 Ibid.

20 Spiegel et al. 2004.

21 Allan & Czeisler 1994.

22 Goichot et al. 1992.

23 Jung et al. 2011.

24 During cellular metabolism, the reduction of molecular oxygen (O_2) to water (H_2O) can lead to the formation of oxygen free radicals in the body. These molecules, which are highly unstable due to their unpaired electrons and consequently raring to pair up with other electrons to form other (potentially radical) molecules, are liable to destroy all or part of the surrounding cell structures and tissue.

25 Blanco et al. 2007.

26 Chang et al. 2008

27 Everson et al. 2005.

28 Kanabrocki et al. 2002.

29 Periasamy et al. 2015.

30 Aubert 2018.

31 Xie et al. 2013.

32 De Vivo et al. 2017; Diering et al. 2017.

33 Wilson & McNaughton 1994.

34 Peigneux et al. 2004.

35 Cf. Squire & Kandel 2005, pp. 35 and 50.

36 Experiments on these two types of memory were conducted in 2002 and 2007 (Fischer et al. 2002; Gais et al. 2007; Walker et al. 2002). The subjects who got to sleep retained the previous day's exercises far better

than those who were kept from sleeping.

37 Plihal & Born 1997.

38 Wagner et al. 2001.

39 Wagner et al. 2005.

40 Benedict et al. 2009.

41 Diekelmann et al. 2008.

42 Walker 2018, p. 205.

43 Stickgold et al. 2001, and Walker 2018, p. 214.

44 Walker 2018, pp. 217–218.

45 Raskind et al. 2003.

46 Walker 2018, p. 212.

47 Van der Helm et al. 2010.

48 Gujar et al. 2011.

49 Walker 2018, p. 226.

50 Arnulf et al. 2014.

51 Vallat et al. 2017.

52 Suzuki et al. 2004.

3. The Sandman's Debtors

1 NSF 2009.

2 Sleep diaries record bedtimes, waking hours, nap times, nighttime and daytime sleep durations, and frequency and duration of night awakenings.

3 Knutson et al. 2010.

4 The SFRMS is a French learned society, founded in 1985, that brings together doctors, researchers, and health professionals to study the mechanisms of sleep, wakefulness, and sleep disorders. It is also responsible for the accreditation of sleep centres in France and helps train clinicians specialising in sleep medicine.

5 INSV 2012.

6 Léger et al. 2012.

7 INSV 2018.

8 Léger et al. 2011.

9 Mumford 1934, p. 14.

10 Paquot 2008, p. 25.

11 Mumford 1934, p. 17.

12 National Sleep Foundation 2013.

13 Ryu et al. 2011.

14 Rowshan Ravan et al. 2010.

15 Hublin et al. 2001.

16 Vela et al. 2012.

17 Garbarino et al. 2016.

18 Ohayon & Milesi 2016.

19 INSV 2014.

20 Montagner 2009.

21 Ibid. (citing Koch et al. 1987).

22 Ibid.

23 Touchette et al. 2007.

24 INSV 2011.

25 INSV 2014.

26 There are plenty of good books about sleep disorders, so this one will not go into the subject at length. I recommend consulting

Billiard & Dauvilliers 2011
and Léger 2017.

27 Dejours 1998.

28 Cited in Vincent 2018.

29 Dejours 2016, p. 256.

30 According to the French
Labour Code (Article
L. 3122-29), 'night work'
means any work performed
between 9.00 pm and 6.00
am for at least three hours
and at least twice a week.

31 http://dares.travail-emploi.
gouv.fr/IMG/pdf/2014-062.
pdf.

32 Ibid.

33 Lee, McCann & Messenger
2007, p. 98.

34 Hamermesh & Stancanelli
2015.

35 Lee, McCann & Messenger
2007, p. 98.

36 Thibault 2014, p. 19; see also
Thibault 2013, p. 167.

37 Roughly 8 per cent and 3.6
per cent more for regular
and occasional night work,
respectively (http://dares.
travail-emploi.gouv.fr/IMG/
pdf/2014-062.pdf).

38 Gamble et al. 2011.

39 https://www.inrs.fr/risques/
travail-de-nuit-et-travail-
poste/donnees-generales-et-

exposition-aux-risques.html

40 Fernandez, Gatounes,
Herbain & Vallejo 2003, p. 100.

41 Ibid., pp. 107–117.

42 Thibault 2013, p. 169.

43 Åkerstedt et al. 2013, p. 29.

44 Ibid., p. 53.

45 Brygo 2018.

46 SFMT 2012; cf. also pp. 125 et
seq.

47 Straif et al. 2007; cf. also http://
www.chu-rouen.fr/sfmt/
autres/Recommandations_
Argumentaire_Version_
juin_2012.pdf.

4. How Deep in Debt Are We?

1 Garbarino et al. 2016 (for
drowsy driving); Faraut et al.
2012 (for high blood pressure);
Zhai et al. 2015 and Dickinson
et al. 2018 (for anxiety and
depression); Bayon et al. 2014
(for obesity); Gangwish et al.
2007 (for diabetes); Ferrie et
al. 2007 and Cappuccio et
al. 2011 (for cardiovascular
events).

2 Klauer et al. 2006. Near
accidents include unintended
lane departures and getting
back into your lane in the

nick of time, or braking suddenly after a driving error.

3 NCSDR/NHTSA 1997.

4 Åkerstedt et al. 2013, p. 21.

5 Ibid.

6 Ibid.

7 Banks & Dinges 2007.

8 Fröberg 1977.

9 Lavie 1989.

10 Faraut et al. 2011; cf. also the editorial in Lange & Born 2011.

11 Edwards et al. 2008.

12 NSF 2015.

13 Onen et al. 2001; Tiede et al. 2010.

14 Faraut et al. 2015a.

15 Roehrs et al. 2012.

16 Work-related musculoskeletal disorders (WMSDs), the most rapidly spreading pathologies of physical 'overload' in Western countries, affect 500,000 people in France (Dejours 2016, p. 258).

17 Baum et al. 2014.

18 Dahl 2008.

19 Kim & Lee 2018.

20 Zhai et al. 2015.

21 INPES 2005.

22 Ibid.

23 Adrien 2002.

24 Vogel et al. 1975.

25 Prather et al. 2015.

26 Cohen et al. 2009.

27 Patel et al. 2012.

28 Irwin et al. 1996.

29 Faraut et al. 2013.

30 Lange et al. 2010; Besedovsky et al. 2017.

31 Lange et al. 2003.

32 Spiegel et al. 2002.

33 Everson 1993.

34 Jonathan Crary, a professor of modern art and theory at Columbia University, writes insightfully in *24/7: late capitalism and the ends of sleep*, 'One of the many reasons human cultures have long associated sleep with death is that they each demonstrate the continuity of the world in our absence' (2013).

35 To cite just a few since the turn of the millennium: Heslop et al. 2002; Ayas et al. 2003; Gottlieb et al. 2005; Ferrie et al. 2007; King 2008; Cappuccio et al. 2011.

36 Ayas et al. 2003.

37 Heslop et al. 2002.

38 King et al. 2008.

39 Faraut et al. 2012.

40 Cappuccio et al. 2011.

41 Scheer et al. 2009.

42 Markwald et al. 2013.

43 Spiegel et al. 1999;

Nedeltcheva et al. 2009.

44 Schmid et al. 2011.

45 The overall prevalence of diabetes — type-2 diabetes in 90 per cent of cases — was estimated at 4.6 per cent of the French population in 2011. But this figure is an underestimate because about 20 per cent of diabetes cases in adults aged 18–74 go undiagnosed. The highest incidence of this disease, which usually appears after the age of 40 (though most frequently diagnosed around the age of 65), is among people aged 75–79 (20 per cent of men and 14 per cent of women). However, since lifestyle is the principal risk factor, type-2 diabetes can also affect adolescents and even children.

46 Boudjeltia & Faraut et al. 2011.

47 Zhong et al. 2005 (one sleepless night); Irwin et al. 1999 (a night of four hours' sleep from 11.00 pm to 3.00 am); Lusardi et al. 1996 (a night of three hours' sleep from 2.00 to 5.00 am); Spiegel et al. 1999 (six nights of four hours' sleep).

48 Ramachandruni et al. 2004.

49 Lusardi et al. 1996.

5. Pocket Medicine

1   The greater the number of awakenings, the more they induce the hallucinations and visions during the stages of falling asleep and periods of so-called 'wake after sleep onset' — in other words, during the borderline state between consciousness and unconsciousness, to which appetitive nappers might be particularly sensitive.

2   Evans et al. 1977.

3   Some sophrologists have made this a speciality in their practice.

4   Debellemaniere et al. 2018.

5   Cordi et al. 2014.

6   Milner et al. 2006 (data obtained by EEG).

7   Lopez-Minguez et al. 2017.

8   Dinges 1992.

9   Lampl & Johnson 2011.

10 Brescianini et al. 2011; Touchette et al. 2013.

11 Weissbluth 1995; Komada et al. 2012.

12 Thorleifsdottir et al. 2002.

13 INSV 2018.

14 Jakubowski et al. 2017.

15 Lavie 1986.

16 Lowden et al. 2004.

17 INSV 2010 (for France); Picarsic et al. 2008 (for the US).

18 Ficca et al. 2010.

19 Mindell et al. 2013.

20 NSF 2013.

21 Bouscoulet et al. 2008.

22 INSV's 2014 study for this population segment.

23 INSV's 2012 study for this population segment.

6. The Art of the Siesta

1 Zhao et al. 2010 (using polysomnographic analysis in particular).

2 Roach et al. 2018.

3 Others hold a pen in their hand, but sleeping in an upright position will suffice, for the tilting of your head as you fall asleep is bound to give you a start.

4 Dalí 1992.

5 Tietzel & Lack 2002. This sort of 'micro-nap' does provide some mental relaxation, however.

6 Milner & Cote 2009.

7 The indicated durations of the various stages of sleep during a nap are rough approximations and will naturally vary according to the sleeper and their nap episodes.

8 Hayashi et al. 2005.

9 Brooks & Lack 2006.

10 Expect to get 20–30 minutes of deep slow-wave sleep during a one-hour nap in the early afternoon, and 45 minutes during a 90-minute nap (Mednick et al. 2003; Nishida & Walker 2007, with a population sample aged 18–30).

11 Vgontzas et al. 2007.

12 Schweitzer et al. 2006

13 Tsapanou et al. 2016.

14 Cross et al. 2015.

15 Gulia & Kumar 2018.

16 Lin et al. 2018.

17 Cao et al. 2014.

18 Yamada et al. 2015.

19 Mantua & Spencer 2015. Likewise, in a recent study by the Hôtel-Dieu Sleep Centre, we found a higher level of inflammation in AIDS patients who frequently take long naps (Faraut et al. 2018).

20 Naitoh 1981.

21 As mentioned, besides being

less effective, late-afternoon and evening naps can impair the quality of subsequent night sleep.

22 Lavie & Weler 1989.

23 Suzuki et al. 2004.

24 Lahl et al. 2008.

25 Nishida et al. 2007.

26 Genzel et al. 2014.

27 Morita et al. 2016.

28 Mednick et al. 2003.

29 Seehagen et al. 2015.

30 Horvath et al. 2015.

31 Kurdziel et al. 2013.

32 Zhao et al. 2010.

33 Kaida et al. 2007.

34 The exercise consists in finding an associative link between seemingly unrelated words, e.g. 'sweet' for the words 'sixteen', 'cookies', and 'heart'.

35 Cai et al. 2009.

36 Born et al. 1999.

37 If coffee prevents you from falling asleep, since its psychostimulant effect is always accompanied by the release of adrenaline and increased heart rate, simply have your cup after the nap.

38 Cajochen 2007; Lockley et al. 2006; Revell et al. 2006; West et al. 2011.

39 Faraut et al. 2020.

40 Hayashi et al. 2003.

41 Jaadane et al. 2017.

7. A Therapeutic Stroll

1 Sagaspe et al. 2007.

2 Garbarino et al. 2004.

3 Faraut et al. 2015a.

4 Theadom et al. 2015.

5 Takahashi et al. 2009.

6 Alexandre et al. 2017.

7 Meerlo et al. 2015.

8 Kaida et al. 2007; Yoon et al. 2003.

9 Dimitrov et al. 2009.

10 Vgontzas et al. 2007.

11 Faraut et al. 2011 and 2015b.

12 Zaregarizi et al. 2007.

13 Gronfier et al. 1998; see also Besedovsky et al. 2017.

14 Naska et al. 2007.

15 Besedovsky et al. 2012.

16 Boudjeltia et al. 2008; Kerkhofs et al. 2007; Boudjeltia & Faraut et al. 2011; Faraut et al. 2011.

17 Loimaala et al. 2006.

18 Ruggiero et al. 2007.

19 Faraut et al. 2011.

20 Dinges et al. 1994.

21 Lasselin et al. 2015.

22 Faraut et al. 2011.

23 Vgontzas et al. 2007. It

should be noted that, in addition to the anti-inflammatory effect of a nap, it also has an analgesic effect insofar as interleukin-6 is a pain mediator as well (Zhou et al. 2016).

24 Kimura & Kishimoto 2010.

25 Faraut et al. 2015b.

26 Mantua & Spencer 2015.

27 Irwin et al. 2016.

28 Trichopoulos et al. 1987.

29 Naska et al. 2007.

30 Sayón-Orea et al. 2013.

31 Kang et al. 2017.

32 Yamada et al. 2015.